INTERNATIONAL SERIES OF
MONOGRAPHS ON COMPUTER SCIENCE

THE INTERNATIONAL SERIES OF
MONOGRAPHS ON COMPUTER SCIENCE

1. *The design and analysis of coalesced hashing* Jeffrey S. Vitter and Wen-chin Chen
2. *Initial computability, algebraic specifications, and partial algebras* Horst Reichel
3. *Art gallery theorems and algorithms* Joseph O'Rourke
4. *The combinatorics of network reliability* Charles J. Colbourn
5. *Discrete relaxation techniques* T. Henderson
6. *Computable set theory* D. Cantone, A. Ferro, and E. Omodeo
7. *Programming in Martin-Löf's type theory: an introduction* Bengt Nordström, Kent Petersson, and Jan M. Smith
8. *Nonlinear optimization: complexity issues* Stephen A. Vavasis
9. *Derivation and validation of software metrics* Martin Shepperd and Darrel Ince
10. *Automated deduction in multiple-valued logics* Reiner Hähnle

Automated Deduction in Multiple-valued Logics

REINER HÄHNLE

University of Karlsruhe

CLARENDON PRESS · OXFORD

1993

Oxford University Press, Walton Street, Oxford OX2 6DP

Oxford New York Toronto
Delhi Bombay Calcutta Madras Karachi
Kuala Lumpur Singapore Hong Kong Tokyo
Nairobi Dar es Salaam Cape Town
Melbourne Auckland Madrid
and associated companies in
Berlin Ibadan

Oxford is a trade mark of Oxford University Press

Published in the United States by
Oxford University Press Inc., New York

A catalogue record for this book is available from the British Library

Library of Congress Cataloging in Publication Data
(data available)

ISBN 0–19–853989–4

Typeset by the author
Printed in Great Britain by
Bookcraft (Bath) Ltd,
Midsomer Norton, Avon

ACKNOWLEDGEMENTS

This book is a revised version of my PhD thesis, submitted to the Department of Computer Science, University of Karlsruhe in May 1991.

Many people have helped and encouraged me in many ways while this book and the PhD thesis from which it emerged was being prepared.

My sincerest thanks go to—my parents for all their love and support over the years; Gabi for being a true friend in all seasons; Uli for his friendship, his humour and his delicious ice creams; Prof. P. Schmitt for being the most excellent advisor one could wish for and for suggesting that I investigate many-valued theorem proving in the first place; Prof. D. Mundici for being co-referee of my thesis, for great encouragement, for many stimulating discussions, and for providing important contacts; Prof. W. Menzel for waking my interest in Theoretical Computer Science ten years ago; the Department of Computer Science at the University of Karlsruhe for granting that my thesis could be written in English, otherwise this book would not have been produced; B. Beckert, S. Gerberding, and W. Kernig for spending their time and expertise in the construction of $_3TAP$ far beyond their duties; IBM Germany for providing the financial framework of the TCG project; M. D'Agostino, J. Dix, I. Gent, M. Kummer, J. Lu, B. Ludäscher, N. Murray, J. Posegga, W. Reif, E. Rosenthal, Z. Stachniak for valuable discussions and comments at one time or another; Prof. D. Gabbay, Imperial College, London, for communicating my work to OUP; a 'retired English Gentleman without hair' who prefers not be mentioned for removing grammatical mistakes and improving the style of my English; and, finally, to Groucho Marx, Monty Python, David Lodge, Kingsley Amis, and Ernst Lubitsch for cheering me up more than once. I am grateful to all of you.

Any mistakes, errors and shortcomings still present are, of course, my own and the above mentioned people and institutions are not to be blamed for them.

I gratefully dedicate this book to all my teachers, be it in life or in science.

Karlsruhe   R. H.
June 1993

CONTENTS

1	**Introduction**	1
2	**Preliminaries**	4
2.1	Abstract algebras	4
2.2	Syntax and semantics	5
2.3	Examples of multiple-valued logics	11
	2.3.1 The logics of Kleene and of Rosser and Turquette	12
	2.3.2 Łukasiewicz Logics	13
	2.3.3 Post logics	14
3	**The Logical Basis: Signed Analytic Tableaux**	15
	Introduction to chapters 3 to 6	15
3.1	Signed tableaux for classical logic	16
3.2	On the rôle of signs in analytic tableaux	22
3.3	Multiple-valued extension of tableau systems	23
3.4	Discussion	29
4	**A New Technique: Truth Value Sets as Signs**	32
4.1	Sets as signs	32
4.2	Soundness	40
4.3	Completeness	42
4.4	Size of proof trees	47
4.5	Function minimization	50
5	**Uniform Notation Regained: Regular Logics**	55
5.1	Primary multiple-valued connectives	55
5.2	Regular logics	59
5.3	On the scope of regular logics	68
5.4	First-order multiple-valued logics	71
5.5	Extensions	77

6   Beyond Tableaux   81

6.1   Lemma generation—asymmetric rules—analytic cut   81
    6.1.1   Lemma generation and asymmetric rules   81
    6.1.2   KE systems and the principle of bivalence   83
    6.1.3   Lemma generation in multiple-valued logics   85
    6.1.4   The principle of multivalence   87

6.2   Tableaux as integer programming problems   89
    6.2.1   Mixed integer programming   89
    6.2.2   Tableau proofs with constraints   91
    6.2.3   Example   94
    6.2.4   Complexity of multiple-valued logics   95
    6.2.5   A reduction from multiple-valued deduction to
        bMIP   96
        6.2.5.1   *Negation*   98
    6.2.6   MIP formulation of $\mathcal{L}_\omega$ 99
    6.2.7   Outlook   101

6.3   Other inference systems   102
    6.3.1   Techniques common to tableau-like calculi   102
        6.3.1.1   *Proof representation*   102
        6.3.1.2   *Free variables*   102
    6.3.2   Decision diagrams   103
    6.3.3   Extending multiple-valued dissolution   104
    6.3.4   Resolution   105

6.4   Evaluation   107
    6.4.1   Evaluating the criteria from the introduction   107
    6.4.2   Some experimental results   108
    6.4.3   Outlook   109

7   Applications   110

7.1   Overview   110

7.2   Applications of a theoretical nature   112
    7.2.1   Independence proofs in Hilbert systems   112
    7.2.2   Relevance logic: A special case   113
    7.2.3   Suchoń's tableau system: another special case   115
    7.2.4   Improvement of an **S5**-implementation by Ca-
        ferra and Zabel   116

7.3   Applications of a practical nature   117
    7.3.1   Interval arithmetic   117
    7.3.2   Hardware verification   117

8 A History of Multiple-Valued Theorem Proving   120

Introduction   120

8.1 Resolution-based systems   121
    8.1.1 Morgan's resolution-based system   121
    8.1.2 Schmitt's resolution-based system   125
    8.1.3 Stachniak's resolution logics   127
    8.1.4 Resolution-based systems for fuzzy logic   133
            8.1.4.1 *Lee and Chang's resolution-based system*   134
            8.1.4.2 *Extended Post logics*   134
            8.1.4.3 *ω+1-valued Post logics*   136
    8.1.5 Paraconsistent logics   138

8.2 Other approaches   139
    8.2.1 Decision diagrams   139
            8.2.1.1 *Binary decision diagrams*   139
            8.2.1.2 *n-ary decision diagrams*   140
            8.2.1.3 *Multiple-valued deduction as a unification problem*   142
    8.2.2 Approaches based on tableaux and Gentzen calculi   143
    8.2.3 Path dissolution by Murray and Rosenthal   144
    8.2.4 Beavers' approach to Łukasiewicz logic   147
    8.2.5 Mellouli's three-valued extension of Plaisted's modified problem reduction format   148
    8.2.6 General frameworks   151
            8.2.6.1 *AUTOLOGIC*   151
            8.2.6.2 *Labelled deductive systems*   151

8.3 Discussion   152

9 Conclusion   154

References   156

Index   169

1

INTRODUCTION

La dernière chose qu'on trouve en faisant un ouvrage est de savoir celle qu'il faut mettre la première.
— Blaise Pascal, *Pensées*

This book is, to the best of my knowledge, the first monograph exclusively devoted to automated theorem proving in many-valued[*] logics. There are several books on automated theorem proving, for instance Chang and Lee (1973), Wos *et al.* (1984), Bibel (1987), Wos (1988), Fitting (1990b), and even more on many-valued logic[†] (Rosser and Turquette, 1952; Ackermann, 1967; Rescher, 1969; Dunn and Epstein, 1977; Gottwald, 1989; Bolc and Borowik, 1992). There are also books on automated theorem proving in other non-classical logics, for example, Wallen (1990). In general, the interest in theorem proving in various non-classical logics, triggered mainly by new applications in computer science,[‡] has grown vastly in the last ten years. One can easily find dozens of references on theorem proving in intuitionistic, modal, linear, conditional, non-monotonic and temporal logic, but not so on theorem proving in many-valued logic.

The reasons for this situation are, first, the widespread opinion that many-valued logic is not of much use, except for the pure mathematician, a view expressed, for example, in Urquhart's article on many-valued logic in the *Handbook of Philosophical Logic* (Urquhart, 1986). Indeed, a brief look into the Proceedings of the field's main conference,[§] shows that the major interests are in pure mathematics (algebra, spectral theory) and in the actual implementation of switching circuits. The second reason for the apparent lack of interest in computational many-valued logic is the seeming heterogeneity of many-valued logics[¶] and conceptual opaqueness[‖]

---

[*]In the literature the terms *multiple-valued*, *many-valued* and *multi-valued* logic are used interchangeably, which is also what we will do in this work.

[†]There is even a conference series devoted to the subject, see below.

[‡]There are at least two conference series devoted to logic and computer science. A good deal of the papers have non-classical logics as their subject.

[§]The *International Symposium on Multiple-Valued Logic*; the proceedings are published by IEEE Press.

[¶]Urquhart shows that it is difficult to characterize what exactly the term *many-valued logic* means.

[‖]'Everyone can understand $\{t, f\}$-valuations, but few—even the creators of the subject—can understand many-valued truth tables' (Scott, 1976) which makes it hard to compare different systems against each other and also hinders the development of general proof procedures. Thirdly, it has been argued that many-valued logic suffers from a lack of convincing applications.* In Chapter 7 we show that, at least since recently, this is no longer true. One of the results, however, will be that these applications involve rather different kinds of many-valued logics.

It is therefore not surprising that activities so far in many-valued theorem proving have almost always been restricted to small and short-term projects, and researchers in many cases do not even know of related work done by their colleagues. Also, with few recent exceptions, only specialized logics or restricted classes of logics have been considered for automated theorem proving.

Consequently, one of the contributions of this work is to give the (to the best of our knowledge) most comprehensive overview to date on activities in many-valued theorem proving. Chapter 8 gives an extensive historical account of the subject.

The best argument for doing further research in the subject, however, is to show that it can actually be done, not only in principle, but also efficiently, provided the appropriate theoretical tools are chosen. Starting out from a straightforward solution that uses semantic tableaux (Chapter 3), we develop a notion we have coined **sets-as-signs**, and which is henceforth used to manage an efficient representation of the many-valued computational search space (Chapters 4, 5). Once we have the tool of sets-as-signs we see that it is not limited solely to application on semantic tableaux, but can be used to modify virtually any known inference method to handle many-valued logics. We show this for selected examples (Chapter 6). Therefore, we see the present work as a foundation for many-valued theorem proving.

Before we start with the formalities, let us set up for later use a catalogue of properties which a framework that can serve as a foundation for many-valued theorem proving should satisfy.

- **Wide applicability** For a wide range of logics, including very different kinds, propositional as well as first-order, a theorem proving system must be schematically derivable.
- **Flexibility** A proof system for a new logic can be derived without having to give new proofs for soundness and completeness theorems, when the semantics are given in some standard way.
- **Easy adaptability** Switching to a different logic must be possible without having to redesign the whole prover. Preferably it should be done interactively by the user with a few simple instructions with the necessary changes being computed automatically.

*Urquhart's main argument is that many-valued logic fails to solve the problems that it was designed for by its inventors. Indeed, even in recent overviews such as (Gottwald, 1989), convincing applications are hardly to be found.

- **Performance** The derived theorem provers are to be reasonably efficient with respect to the complexity of the given logic and in comparison with theorem provers for classical logic.
- **Closeness to classical versions** A proof system for many-valued logic should not be dramatically different from a classical version of it, so that techniques of implementation and for improving performance known for the classical case can be reused or easily adopted.

This catalogue will be evaluated in Section 6.4.1.

Since this book emerged from a PhD thesis, naturally it does not have an introductory character, but, rather, is intended for the active researcher. On the other hand, an attempt has been made to make it self-contained and to be generous with the references.

The essential ideas presented in Chapters 4 to 6 appeared in Hähnle (1990b), Hähnle (1991), Hähnle (1992a), Hähnle (1993b), but the text was considerably revised and extended for the present purpose. Some of the material in Chapter 7 appeared in Hähnle (1993c), Hähnle *et al.* (1992); Chapter 8 has not been published before.

PRELIMINARIES

Gebt Ihr ein Stück, so gebt es gleich in Stücken!
Solch ein Ragout, es muß Euch glücken;
Leicht ist es vorgelegt, so leicht als ausgedacht.
Was hilft's, wenn Ihr ein Ganzes dargebracht?
— Johann Wolfgang von Goethe, *Faust*

In this chapter we collect some definitions that we will use frequently in this book and introduce some special notation. Readers who are familiar with formal logic and automated theorem proving can skip this chapter in a first reading.

2.1 Abstract algebras

In order to specify the syntax and semantics of the logics we are going to discuss we use a slightly more general approach than is found in most logic textbooks, namely the concept of an *abstract algebra*.* We borrowed most of the material in this and the following section from Carnielli (1987), Cohn (1981) and Rasiowa (1974). Unlike the latter author, our primary reason for the use of abstract algebras is not the aim of an algebraic treatment of logic, but merely to simplify notation: in many cases we do not want to be specific about the particular connectives or semantics involved. So, following Carnielli (1987), abstract algebras are a convenient device that allows a more general treatment here.

Definition 2.1. (Abstract algebra) *Let A be a non-empty set and $\Omega = (\omega_i)_{i \in \mathbb{N}}$ an indexed family of finitary operators on A. Assume that we have a function α giving the arity of each operator, thus $\omega : A^{\alpha(\omega)} \to A$ for each $\omega \in \Omega$. Call $\mathbf{A} = (A, \Omega)$ an* **abstract algebra** *on A and the sequence $\langle \alpha(\omega_1), \alpha(\omega_2), \ldots \rangle$ its* **similarity type**. *The set A is called the* **universe** *of $\mathbf{A}$. In cases where only finitely many operators are present, we write*

$$(A, \omega_1, \ldots, \omega_m)$$

and $\langle \alpha(\omega_1), \ldots, \alpha(\omega_m) \rangle$. Two algebras that share the same similarity type are called **similar**. *Call $\mathbf{A}_0$ a* **subalgebra** *of $\mathbf{A} = (A, (\omega_i)_{i \in \mathbb{N}})$, iff $\emptyset \neq A_0 \subseteq A$ and A_0 is closed under the operators of $\mathbf{A}$ restricted to A_0 that*

---

*Some authors, for example Cohn (1981), speak of Ω-*algebras*.

*is under $(\omega_i \restriction_{A_0})_{i \in \mathbb{N}}$. A non-empty set A_0 generates a subalgebra $\mathbf{B}$ of $\mathbf{A}$ iff $\mathbf{B}$ is the least subalgebra of $\mathbf{A}$ containing A_0. A set A_0 **generates** $\mathbf{A}$ iff the subalgebra of $\mathbf{A}$ that is generated by A_0 is $\mathbf{A}$ itself.*

Definition 2.2. (Homomorphism) *A mapping $h : \mathbf{A} \to \mathbf{B}$ between similar abstract algebras is called a **homomorphism** from $\mathbf{A}$ to $\mathbf{B}$ iff for all $n \in \mathbb{N}$, $a_1, \ldots, a_n \in A$, $\omega_i^{\mathbf{A}} \in \mathbf{A}, \omega_i^{\mathbf{B}} \in \mathbf{B}$ with $\alpha(\omega_i) = n$ we have*

$$h(\omega_i^{\mathbf{A}}(a_1, \ldots, a_n)) = \omega_i^{\mathbf{B}}(h(a_1), \ldots, h(a_n))$$

Definition 2.3. (Free algebra) *An algebra $\mathbf{A}$ in a class $\mathcal{K}$ of similar algebras is called **free** in $\mathcal{K}$ iff it has a set of generators A_0, such that for all algebras $\mathbf{B} \in \mathcal{K}$ and mappings $f : A_0 \to B$, f can be extended to a homomorphism from $\mathbf{A}$ into $\mathbf{B}$. In this case we say that A_0 **freely generates** $\mathbf{A}$ in $\mathcal{K}$.*

Proposition 2.4. *Any mapping g from the generators of a free algebra $\mathbf{A}$ in $\mathcal{K}$ into any $\mathbf{B} \in \mathcal{K}$ can be extended uniquely to a homomorphism h from $\mathbf{A}$ into $\mathbf{B}$.*

Proof. See Rasiowa (1974).

2.2 Syntax and semantics

The familiar notions of language, formula, and so on are now restated in terms of abstract algebras. For most purposes it suffices to retain the naïve interpretations; however, in order be able to write down precisely the proofs given in Chapter 4 it will be easier to make reference to the following definitions.

The main idea is to use abstract algebras to represent both syntax and semantics. We will take the free term algebra over propositional variables to represent a formal language and we will take a characterization of the algebra of connectives as their semantics. The specification of the latter can then be done with one of the usual methods, for example equationally (Rasiowa, 1974), lattice theoretically (Balbes and Dwinger, 1974), with consequence relations (Wójcicki, 1988), by recursive definition, or simply by explicitly specifying the value of the finite operators in tabular form. The latter method is generally known as *truth table semantics* and it is also the one we will use frequently throughout this book.

Definition 2.5. (Propositional formula) *Consider a free* algebra $\mathbf{L}$ with operations $F_1, \ldots, F_r$, generated by $L_0 = \{p_i | i \in \mathbb{N}\}$. Then we call the members of L_0 **propositional variables** or **atomic formulas** and the members of $\{F_1, \ldots, F_r\}$ **propositional connectives**. Define*

---

* *Free* means always free in the class of similar algebras.

$$
\begin{aligned}
L_{i+1} &= L_i \cup \\
&\quad \{F_j(\phi_1,\ldots,\phi_{\alpha(F_j)}) \mid \phi_1,\ldots,\phi_{\alpha(F_j)} \in L_i, F_j \in \{F_1,\ldots,F_r\}\} \\
L &= \bigcup_{i \in \mathbb{N}} L_i
\end{aligned}
$$

and call L_i the **formulas of depth** i *and L the* **propositional L-formulas**.

We note that the algebra $(L, F_1, \ldots, F_r)$ is well-defined and freely generated by L_0. Thus the following definition is justified.

Definition 2.6. (Propositional language) *The abstract algebra* $\mathbf{L} = (L, F_1, \ldots, F_r)$, *defined as above, is called* **propositional language L**.

Definition 2.7. (Matrix, truth value) *Let* $\mathbf{L}$ *be a propositional language and* $A = (N, f_1, \ldots, f_r)$ *(for a finite set N) be an algebra similar to it. Then we call the algebra* $\mathbf{A} = (N, d_1, \ldots, d_k, f_1, \ldots, f_r)$, *where $k \leq |N|$ and all d_i are 0-ary operators on N, a* **propositional matrix** *for* $\mathbf{L}$. *The set N is called the* **truth value set** *of* $\mathbf{A}$, *while* $D \doteq \{d_1, \ldots, d_k\} \subseteq N$ *is the set of* **designated truth values**. *We denote the cardinality of the set N (resp., D) with n (resp., d).*

The designated truth values support the validity of a statement. In the familiar propositional two-valued case one chooses the two-element Boolean algebra as a matrix.

In our exposition we usually take equidistant numbers from the rational interval $[0, 1]$ to denote the set of truth values and the larger k elements in this set as the designated truth values. Therefore, typically we would have $N = \{0, \frac{1}{n-1}, \ldots, \frac{n-2}{n-1}, 1\}$, $D = \{\frac{n-k}{n-1}, \frac{n-k+1}{n-1}, \ldots, \frac{n-2}{n-1}, 1\}$. When the value of D is obvious or is not of concern we omit D from the definition of a matrix $\mathbf{A}$. Let us further abbreviate $\{0, 1\}$ with $\mathbf{2}$, $\{0, \frac{1}{2}, 1\}$ with $\mathbf{3}$, and generally, $\{0, \frac{1}{n-1}, \ldots, \frac{n-2}{n-1}, 1\}$ with $\mathbf{n}$.

Definition 2.8. (Propositional logic) *Let* $\mathbf{L}$ *be a propositional language and* $\mathbf{A}$ *a propositional matrix for it. Then the pair* $\mathcal{L} = (\mathbf{L}, \mathbf{A})$ *is called* **n-valued propositional logic** *with designated truth values D.*

As in the classical case, a valuation simply assigns a truth value to each atomic formula of the language.

Definition 2.9. (Valuation) *Let* $\mathcal{L}$ *be a propositional logic. A* **propositional valuation** *is a mapping* $v : L_0 \to N$.

Since any valuation v can be uniquely extended to a homomorphism $v' : \mathbf{L} \to \mathbf{A}$ (this follows from Proposition 2.4) we often speak of v', when only v is explicitly defined.

Definition 2.10. (Satisfiability, model, tautology) *Let $\mathcal{L}$ be a propositional logic. A formula $\phi \in L$ is* **propositionally satisfiable** *iff $v(\phi) \in D$ for some $\mathcal{L}$-valuation v. A set of formulas is satisfiable iff some $\mathcal{L}$-valuation simultaneously satisfies each of its members. The valuation that satisfies a formula ϕ is called a* **($\mathcal{L}$-)model** *for ϕ. If a formula ϕ is satisfied by any valuation it is called* **propositional ($\mathcal{L}$-)tautology**.

All of the propositional operators in classical logic can be generalized to many-valued logics. The problem is that, in general, there is more than one way to do this. Consider, for example, the connectives $\neg$ and $\sim$ defined in Tables 2.1 and 2.2 on pp. 12 and 13. Both can serve as a kind of three-valued negation. We introduce the following notions in order to be able to compare different many-valued versions of classical propositional connectives.

Definition 2.11. (Generalized, weak, strong connective) *Let F be a k-ary classical propositional connective and $\mathcal{L}$ an n-valued propositional logic whose language contains the k-ary operator symbol F'. Then F' is an n-***valued generalization** *of F iff $F' \upharpoonright_2 = F$.*

Let F, G be two n-valued propositional connectives in $\mathcal{L}$ with the same arity k. We call F **weaker than** *G (G* **stronger than** *F) iff for all $\mathcal{L}$-valuations v, $\phi_i \in \mathbf{L}$:*

$$v(F(\phi_1, \ldots, \phi_k)) \geq v(G(\phi_1, \ldots, \phi_k))$$

Obviously, this defines a partial order on each class of connectives with the same arity.

Now we are going to extend our case to first-order logic. For the most part this can be done in a straightforward manner.

Definition 2.12. (Term algebra) *Let $E = \bigcup_{i \geq 0} E_i$ be a finite or countable set of function symbols, where each E_i contains i-ary function symbols. Let V be a countable set of object variables. Consider the free algebra $\mathbf{T}$ with operations E, generated by V. Define*

$$
\begin{aligned}
T_0 &= V \\
T_{i+1} &= T_i \cup \{f(t_1, \ldots, t_j) \mid t_1, \ldots, t_j \in T_i, f \in E_j, j \in \mathbb{N}\} \\
T &= \bigcup_{i \in \mathbb{N}} T_i
\end{aligned}
$$

and call T the **terms over** *$V \cup E$. We note that $\mathbf{T} = (T, E)$ and call $\mathbf{T}$ the* **free term algebra** *over V.*

Note that T is not empty, since $V \subseteq T$.

Definition 2.13. (First-order formula) *Let $\mathbf{T} = (T, E)$ be a term algebra, $\mathbf{L} = (L, F_1, \ldots, F_r)$ a propositional language, and $P = \bigcup_{i \geq 0} P_i$ a* *non-empty finite or countable set of predicate symbols, where each P_i contains i-ary predicates. Define the set L^1 of* **first-order (L^1-)formulas** *as follows:*

$$
\begin{aligned}
L_0^1 &= \{p(t_1,\ldots,t_j)\mid t_1,\ldots,t_j \in T, p \in P_j, j \in \mathbb{N}\} \\
L_{i+1}^1 &= L_i^1 \cup \\
&\quad \{F_j(\phi_1,\ldots,\phi_{\alpha(F_j)}) \mid \phi_1,\ldots,\phi_{\alpha(F_j)} \in L_i^1, F_j \in \{F_1,\ldots,F_r\}\} \cup \\
&\quad \{(Qy)\phi \mid y \in V, Q \in \{\forall,\exists\}, \phi \in L_i^1\} \\
L^1 &= \bigcup_{i\in\mathbb{N}} L_i^1
\end{aligned}
$$

The formulas in L_0^1 are called **atomic formulas**. *The formulas in the sets L_i^1 are called* **first-order formulas of depth i**. *The abstract algebra $\mathbf{L}^1 = (L^1, F_1,\ldots,F_r)$, where the F_i are defined as in $\mathbf{L}$, will be called* **first-order language $\mathbf{L}^1$ associated with $\mathbf{L}$**.

We note that $\mathbf{L}^1$ is similar to $\mathbf{L}$.

On various occasions we need first-order languages that contain parameters and Skolem functions.

Definition 2.14. (Parameter, Skolem function)   *Let $\mathbf{L}^1$ be a first-order language and* **par** *a countable set of constant symbols disjoint from E_0. If we replace E_0 in Definition 2.12 with $E_0 \cup$ **par** we call the resulting first-order language $\mathbf{L}^{\text{par}}$ and the members of* **par** *parameters. Similarly, if* **sko** $= \bigcup_{i\geq 0}$ **sko**$_i$ *is a countable set of function symbols that contains infinitely many function symbols* **sko**$_i$ *disjoint from E_i for each arity, we construct a first-order language $\mathbf{L}^{\text{sko}}$ by replacing E_i with $E_i \cup$ **sko**$_i$ at each stage of the definition of T. We call the additional symbols* **Skolem function symbols** *and* **Skolem constant symbols**.

We define **free (bound) occurrences** of variables in formulas exactly as in the classical case and refer the reader to any logic textbook, for example see Fitting (1990b). We use the notation $\phi(x)$ if the variable x occurs freely in a formula ϕ. Formulas without occurrences of free variables are called **closed formulas** or **sentences**.

Definition 2.15. (Substitution) *Let $\mathbf{L}^1$ be a first-order language. A* **substitution** *is an arbitrary mapping $\sigma : V \to T$.*

Again, σ can be uniquely extended to $\sigma' : \mathbf{T} \to \mathbf{T}$ and subsequently to $\sigma'' : \mathbf{L}^1 \to \mathbf{L}^1$ in the usual manner, see Fitting (1990b) for details. We denote the result of applying a substitution σ to a formula ϕ with $\phi\sigma$. We assume that the reader is familiar with the standard concepts and results of unification theory. If $\sigma(x) \neq x$ for only finitely many $x \in V$, say for

$x_1, \ldots, x_n$, we denote σ with $\{x_1 \leftarrow t_1, \ldots, x_n \leftarrow t_n\}$ where $\sigma(x_i) = t_i$, $1 \leq i \leq n$.

Definition 2.16. (First-order structure) *Let* $\mathbf{L}^1$ *be any first-order language. We call* $\mathbf{M} = (U, I)$ *a* **first-order structure** *for* $\mathbf{L}^1$, *provided that:*

1. *U is a non-empty set, called the* **domain** *or* **universe** *of* $\mathbf{M}$.
2. *I assigns to each*
   (a) *$f \in E_i$ a mapping $I(f) : U^i \to U$,*
   (b) *$p \in P_i$ a mapping $I(p) : U^i \to N$.*

Definition 2.17. (First-order logic) *Let* $\mathbf{L}^1$ *be a first-order language and* $\mathbf{A}$ *a propositional matrix of the associated propositional language. Then the pair* $\mathcal{L}^1 = (\mathbf{L}^1, \mathbf{A})$ *is called an* n-**valued first-order logic** *with designated truth values* D.

The semantics of quantifiers is not mentioned in the definition above, since it is regarded as fixed in all logics that we will consider.

Definition 2.18. (Assignment, variant) *Let* $\mathcal{L}^1$ *be a first-order logic and* $\mathbf{M}$ *a structure for it. We call any mapping* $\beta : V \to U$ *an* **assignment** *in* $\mathbf{M}$. *Given an assignment* β *we define the* x-**variant of** β *with value* u, β_x^u *as follows:*

$$\beta_x^u(y) = \begin{cases} \beta(y) & y \neq x \\ u & y = x \end{cases} \quad \text{for all } u \in U, x, y \in V$$

Obviously, any assignment $\beta : V \to U$ can be uniquely extended to $\beta' : T \to U$. We will tacitly make use of this fact in the following.

There are several ways in which a (many-valued) valuation function for a given first-order formula, structure, and assignment can be defined. For instance, one could parallel the definitions of formula, language, substitution, and so on, relative to the domain U instead of the set of terms T, then take a propositional valuation for this domain-based language, extend it with appropriate rules for quantifiers, and consider an arbitrary U in the validity definition. This approach was taken by Carnielli (1987) following Smullyan (1968) to ease the handling of generalized quantifiers. Another option, carried out, for example, by Morgan (1976) is to specify a truth value as an argument of the valuation function which now yields a (possibly empty) set of satisfying assignments. The approach that we have taken (along the lines of Fitting (1990b)) is to stay as closely as possible to the usual two-valued definitions, which seems to be most appropriate when no unusual quantifiers are involved. See Section 5.5 for a possible generalization of the truth value semantics.

Definition 2.19. (First-order valuation) *Let* $\mathcal{L}^1$ *be a first-order logic,* $\mathbf{M}$ *a structure for it and* β *an assignment in* $\mathbf{M}$. *The function* $v_\beta : L^1 \to N$

is a **first-order** $\mathcal{L}^1$-**valuation,** *provided the following holds for all* L^1-*formulas:*

1. $v_\beta(p(t_1, \ldots, t_n)) = I(p)(\beta(t_1), \ldots, \beta(t_n))$ *for atomic formulas.*
2. $v_\beta(F(\phi_1, \ldots, \phi_{\alpha(F)})) = f(v_\beta(\phi_1), \ldots, v_\beta(\phi_{\alpha(F)}))$, *where* f *is the operation in* $\mathbf{A}$ *that corresponds to* F *in* $\mathbf{L}^1$.
3. $v_\beta((\forall y)\phi) = \min\{v_{\beta_y^u}(\phi)|u \in U\}$, *where* $\min$ *is interpreted naturally on* N.
4. $v_\beta((\exists y)\phi) = \max\{v_{\beta_y^u}(\phi)|u \in U\}$, *where* $\max$ *is interpreted naturally on* N.

Definition 2.20. (First-order satisfiability, tautology) *Let* $\mathcal{L}^1$ *be an* n-*valued first-order logic. A formula* $\phi \in L^1$ *is called* **first-order satisfiable** *iff there is a structure* $\mathbf{M}$ *for* $\mathcal{L}^1$ *such that for some assignment* β *of* $\mathbf{M}$, $v_\beta(\phi) \in D$. *We say that* ϕ **is true in** $\mathbf{M}$ *iff it is satisfied under all assignments of* $\mathbf{M}$, *it is* **valid** *iff it is true in all structures for* $\mathcal{L}^1$. *Valid* $\mathcal{L}^1$-*sentences are called* **first-order tautologies**.

It is obvious that the truth value of a sentence in a structure $\mathbf{M}$ does not depend on the assignments of $\mathbf{M}$. In this case we usually drop the subscript β from v. Note that, as in the classical case, all propositional tautologies are also first-order tautologies, but not vice versa.

Definition 2.21. (First-order consequence) *Consider a first-order logic* $\mathcal{L}$ *and sentences* $\Phi \subseteq \mathbf{L}, \psi \in \mathbf{L}$. *We say that* ψ *is a* $(\mathcal{L})$-**consequence** *of* Φ, *in symbols* $\Phi \vDash_\mathcal{L} \psi$, *iff for all* $(\mathcal{L})$-*structures* $\mathbf{M}$ $v(\Phi) \subseteq D$ *implies* $v(\psi) \in D$.

Of course, $\emptyset \vDash_\mathcal{L} \psi$ iff ψ is a first-order tautology. In this case we write $\vDash_\mathcal{L} \psi$ instead of $\emptyset \vDash_\mathcal{L} \psi$.

Sometimes it will be convenient to introduce syntactical abbreviations in a language.

Definition 2.22. (Formula schema) *Let* $\mathbf{L}^1$ *be a first-order language and* **fs** *be a countable set of new symbols. Let* $\mathbf{L}^{\mathbf{fs}}$ *be the language that results when the atomic formulas of* $\mathbf{L}^1$ *are extended by* **fs**. *Then we call a member* ψ *of* $\mathbf{L}^{\mathbf{fs}}$ *that contains symbols* $\phi_1, \ldots, \phi_n$ *from* **fs** *a* **formula schema** *of* $\mathbf{L}^1$ *and denote it by* $\psi[\phi_1, \ldots, \phi_n]$.

Valuations can be extended naturally to $\mathbf{L}^{\mathbf{fs}}$ if we treat expressions from **fs** as propositional variables.

Definition 2.23. (Abbreviation, syntactical variant)   *Let* $\mathbf{L}^1$ *be a first-order language and* $\psi[\phi_1, \ldots, \phi_n]$, $\theta[\phi_1, \ldots, \phi_n]$ *formula schemata in* $\mathbf{L}^1$. *We define* $\psi[\phi_1, \ldots, \phi_n]$ *as an* **abbreviation** *for* $\theta[\phi_1, \ldots, \phi_n]$ *by the expression*

$$\psi[\phi_1, \ldots, \phi_n]_{def} = \theta[\phi_1, \ldots, \phi_n].$$

Let **Abbr** *be a set of abbreviations. Two* $\mathbf{L}^1$*-formulas* ρ, τ *are* **syntactical variants** *of each other with respect to* **Abbr** *iff there exists a matching substitution* $\sigma : \mathbf{fs} \to L^1$ *and an abbreviation* $\psi[\phi_1, \ldots, \phi_n]_{def} = \theta[\phi_1, \ldots, \phi_n] \in$ **Abbr** *such that* ρ *and* τ *can be made syntactically identical by replacing zero or more occurrences of* $\psi[\phi_1, \ldots, \phi_n]\sigma$ *in* ρ *with* $\theta[\phi_1, \ldots, \phi_n]\sigma$. *They are called* **syntactically equivalent** *with respect to* **Abbr**, *denoted by* $\rho \Leftrightarrow \tau$ *iff there is a finite set of* $\mathbf{L}^1$*-formulas* $\nu_0, \ldots, \nu_k$ $(k \geq 0)$ *such that* $\nu_0 = \rho, \nu_k = \tau$ *and* ν_i *is a syntactical variant of* ν_{i+1}, $0 \leq i \leq k - 1$.

Definition 2.24. (Literal, clause) *Consider a first-order language* $\mathbf{L}^1$ *whose associated propositional language is given by* $(L, \wedge, \vee, o_1, \ldots, o_r)$ *with similarity type* $\langle 2, 2, 1, \ldots, 1 \rangle$. *A* **literal** *is a formula of the form* $o_{i_1} \cdots o_{i_k} p$, *where* $k \geq 0$ *and* p *is atomic. A* **clause** *is a formula of the form* $l_1 \vee \cdots \vee l_m$, *where* $m \geq 0$ *and the* l_i *are literals. We use the notation* $\bigvee_{i=1}^{m} l_i$ *for clauses. If* $m = 0$ *we speak of the empty clause and denote it by* $\square$. *We stipulate that* $v(\square) = 0$ *for all* v *by definition.*

Definition 2.25. (Negation normal form) *Let* $\mathbf{L}^1$ *be as above, but with only a single unary operator* $\neg$ *present. A formula is said to be in* **negation normal form (NNF)** *if* $\neg$ *only occurs before atomic formulas.*

Definition 2.26. (Conjunctive normal form) $\mathbf{L}^1$ *is as above. A formula is said to be in* **conjunctive normal form (CNF)** *if it is in the form* $C_1 \wedge \cdots \wedge C_s$, *where* $s > 0$ *and the* C_j *are clauses. We use the notation* $\bigwedge_{j=1}^{s} C_j$ *for formulas in CNF.*

Definition 2.27. (Prenex normal form, matrix) $\mathbf{L}^1$ *is as above. A formula* ϕ *is said to be in* **prenex normal form** *if it is of the form* $(Q_1 x_1) \cdots (Q_t x_t) E$, *where* $t \geq 0$, $Q_i \in \{\forall, \exists\}$ *and* E *is quantifier-free. The formula* $E \in \mathbf{L}^1$ *is also called the* **matrix*** *of* ϕ, *which is said to be in* **universal (existential) prenex form** *if all the* Q_i *are of universal (existential) type. A formula is in* **prenex CNF** *if it is in prenex normal form and its matrix is in CNF.*

2.3 Examples of multiple-valued logics

In this section we look at some many-valued logics which will be used several times throughout the rest of the book and are collected here so that they can be referred to quickly. At the same time these logics can serve as examples to illustrate the preceding definitions.

---

*There is no danger of confusion with our earlier notion of matrix. It will always be clear from the context which one is meant.

2.3.1   The logics of Kleene and of Rosser and Turquette

These logics were first introduced in Rosser and Turquette (1952). They bear a close resemblance to the logics used by Kleene (1938); see below.

Definition 2.28. ($\mathcal{L}_M^n$) *Let $\mathcal{L}_M^n$ be a first-order logic defined as follows:*

- *There are no 0-ary predicate symbols.*
- *The associated propositional language $\mathbf{L}_M^n$ is given by the abstract algebra $(L_M^n, \wedge, \vee, \neg, J_0, \ldots, J_1)$ with similarity type $\langle 2, 2, 1, 1, \ldots, 1 \rangle$.*
- *The matrix is $(\mathbf{n}, \frac{n-k}{n-1}, \frac{n-k+1}{n-1}, \ldots, 1, \wedge, \vee, \neg, J_0, \ldots, J_1)$ for some $n > k > 0$, where the operators* are specified as follows:*

   1. *$i \wedge j = \min\{i, j\}$, where $\min$ is interpreted naturally on N.*
   2. *$i \vee j = \max\{i, j\}$, where $\max$ is interpreted naturally on N.*
   3. *$\neg i = 1 - i$.*
   4. *$J_k(i) = \begin{cases} 1 & i = k \\ 0 & i \neq k \end{cases}$ for $k \in N$.*

*We call $\neg$ **strong negation** and the J_i's **truth value assertions**.*

For $n = 3$ we also give the definition of the propositional matrix in tabular form in Table 2.1.

Table 2.1   *Truth tables of propositional connectives in $\mathcal{L}_M^n$ for $n = 3$*

$\wedge$	0	$\frac{1}{2}$	1		$\vee$	0	$\frac{1}{2}$	1			$\neg$			J_0	$J_{\frac{1}{2}}$	J_1
0	0	0	0		0	0	$\frac{1}{2}$	1		0	1		0	1	0	0
$\frac{1}{2}$	0	$\frac{1}{2}$	$\frac{1}{2}$		$\frac{1}{2}$	$\frac{1}{2}$	$\frac{1}{2}$	1		$\frac{1}{2}$	$\frac{1}{2}$		$\frac{1}{2}$	0	1	0
1	0	$\frac{1}{2}$	1		1	1	1	1		1	0		1	0	0	1

It is easy to prove that for each n the propositional part of $\mathcal{L}_M^n$ is functionally complete, i.e. every n-valued connective is definable by the ones that are already present.

Definition 2.29. ($\mathcal{L}_{M+}^n$, **weak connectives, equivalences**)   *Extend $\mathcal{L}_M^n$ with a unary propositional operator $\sim$ and binary operators $\supset$, $\equiv$, $\cong$ whose semantics for $n = 3, D = \{1\}$ are defined according to Table 2.2. In general, the semantic definitions are as follows:*

1. *$\sim i = \begin{cases} 1 & i \in N - D \\ 0 & i \in D \end{cases}$.*

2. *$i \supset j = \begin{cases} 1 & i \in N - D \\ j & i \in D \end{cases}$.*

3. *$i \equiv j = \begin{cases} 1 & (i \in N - D \text{ and } j \in N - D) \text{ or} \\ & (i \in D \text{ and } j \in D) \\ \min\{i, j\} & otherwise \end{cases}$*

---

*Since no confusion can arise, we denote the propositional operators and their semantical interpretations by the same symbols.

Table 2.2 *Truth tables of weak connectives in $\mathcal{L}^3_{M+}$*

	$\sim$
0	1
$\frac{1}{2}$	1
1	0

$\supset$	0	$\frac{1}{2}$	1
0	1	1	1
$\frac{1}{2}$	1	1	1
1	0	$\frac{1}{2}$	1

$\equiv$	0	$\frac{1}{2}$	1
0	1	1	0
$\frac{1}{2}$	1	1	$\frac{1}{2}$
1	0	$\frac{1}{2}$	1

$\cong$	0	$\frac{1}{2}$	1
0	1	0	0
$\frac{1}{2}$	0	1	0
1	0	0	1

$$4.\ i \cong j = \begin{cases} 1 & i = j \\ 0 & i \neq j \end{cases}.$$

These connectives are called **weak negation, weak implication, weak equivalence,** *and* **strong equivalence,** *respectively.** Additionally, we allow 0-ary predicate symbols from now on. We call the resulting logic $\mathcal{L}^n_{M+}$.*

A restriction of $\mathcal{L}^n_{M+}$ to certain connectives yields the family of strong Kleene logics which were introduced by Kleene (1938) in the context of partially defined recursive functions (originally only for the propositional case and $n = 3$, but the generalization is straightforward).

Definition 2.30. ($\mathcal{L}^n_{SKL}$) *Let $\mathcal{L}^n_{SKL}$ be a first-order logic defined as follows:*

- *The associated propositional language $\mathbf{L}^n_{SKL}$ is given by the abstract algebra $(L^n_{SKL}, \wedge, \vee, \supset, \neg)$ with similarity type $\langle 2, 2, 2, 1 \rangle$.*
- *The matrix is $(\mathbf{n}, 1, \wedge, \vee, \supset, \neg)$, where the operators are defined as in $\mathcal{L}^n_{M+}$.*

$\mathcal{L}^n_{SKL}$ *is called n-valued* **strong Kleene logic.**

We note that $\mathcal{L}^n_{SKL}$ is not functionally complete for any n.

2.3.2 *Łukasiewicz Logics*

Definition 2.31. ($\mathcal{L}_3$) *Let $\mathcal{L}_3$ be a three-valued first-order logic according to the following specifications:*

- *The associated propositional language $\mathbf{L}_3$ is given by $(L_3, \supset, \vee, \neg)$ with similarity type $\langle 2, 2, 1 \rangle$.*
- *Let $(3, 1, \supset, \vee, \neg)$ be the matrix for the associated propositional language where the operators $\vee$ and $\neg$ are defined as in $\mathcal{L}^n_{M+}$ and the operator $\supset$ by*

$$i \supset j = \min\{1, 1 - i + j\}$$

This is the three-valued logic named after Łukasiewicz which was defined in his famous paper (Łukasiewicz, 1920).

By merely changing the set of truth values N to $\mathbf{n}$ we obtain the class of **finitely-valued Łukasiewicz logics** denoted $\mathcal{L}_n$, and for the choice $N =$

---

*It is easy to see that these are generalizations of classical negation, implication, and equivalence.

[0, 1] **infinitely valued Łukasiewicz logic** $\mathcal{L}_\omega$ (instead of the rational
unit interval we could as well have taken the unit interval of the reals since
$\mathcal{L}_\omega$ has strong finiteness properties).

2.3.3 *Post logics*

We continue with the definition of syntax and semantics of what is often
called n-valued Post logic.[*]

Definition 2.32. ($\mathcal{L}_P^n$) *We call a first-order logic $\mathcal{L}_P^n$ an n-valued Post*
logic, *whenever*

- *its associated propositional language* $\mathbf{L}_P^n$ *is defined by the algebra*
  $(L_P^n, \wedge, \vee, \sigma)$ *with similarity type* $\langle 2, 2, 1 \rangle$,
- *the matrix* $(\mathbf{n}, 1, \wedge, \vee, \sigma)$ *is defined as usual for* $\wedge$ *and* $\vee$ *and*

$$\sigma(i) = \min\left\{1, i + \frac{1}{n-1}\right\} \text{ for all } i \in \mathbf{n}$$

We will see in Section 5.3 that each Post logic is functionally complete.
This property is mainly due to the σ operator, which can also be imagined
as a *rotate left* operation when the truth table is drawn as usual.[†]

> *"Glad that's over", said Mrs. Elmhurst, uncovering her face. "Now what*
> *comes next? A tableau...?"*
> — Virginia Woolf, *Between the Acts*

---

[*]In honour of Emil Post, who initiated (Post, 1921), independently from Łukasiewicz,
the development of many-valued logics using the logics defined below.

[†]It would be as well to define σ as a *rotate right* operation and this is in fact what
has been done by some authors. Post defined 0 as the designated truth value, with truth
tables changed accordingly, and this seems to have caused some confusion.

3

THE LOGICAL BASIS: SIGNED ANALYTIC TABLEAUX

'... Tableau!'
— Wolfgang von Nibelschütz, *Der Blaue Kammerherr*

Introduction to Chapters 3 to 6

The formal proof system called semantic (or analytic) tableaux was introduced by E. W. Beth (1986) and K. J. J. Hintikka (1955) in the 1950s, its ancestors being Gentzen systems. R. Smullyan (1968) gave a particularly elegant version of tableaux which largely increased their popularity, and most tableaux systems used today are based on the formulation he set out. Semantic tableaux are popular as a pedagogical tool; however, recently, their importance in the field of automated theorem proving has also grown substantially. A common prejudice is that tableau-based provers are hopelessly inefficient. Recent implementations (Oppacher and Suen, 1988; Letz *et al.*, 1991) based on variants of the tableau method, on the other hand, have shown that this is not necessarily true for all time. The reason for the relative inefficiency of tableau-based provers so far may very well be a lack of research effort invested into tableau methods in comparison with the more than two decades of attention which resolution-based approaches enjoyed. But whether tableau-like calculi will provide an alternative to resolution-based systems or not (we will not discuss this issue any further in the present work), they constitute a starting point for developing proof procedures in all kinds of non-classical logics. Once a tableau system has been derived for a logic, it is possible to make use of the insight gained therebye into its search space and to develop a more efficient inference method than pure tableaux, be it a tableau refinement or some variant of resolution. A paradigmatic example of this process constitute the work of Wallen (1990) and Ohlbach (1989) for modal and intuitionistic logics. Our methods are different from Wallen's (since modal logic is different from many-valued logic), but the intention is the same in the present work, namely to provide a computationally adequate representation of satisfiability and validity in those logics.

We will provide a brief introduction to semantic tableaux in Chapter 3. In the same chapter we will sketch a method developed by Surma (1984) and Carnielli (1987) that generalizes signed semantic tableaux to arbitrary finitely-valued logics.

Taking this as a starting point, in Chapter 4 we introduce a generalized notion of signs in tableaux that enables us to speak concisely about the truth values a formula can take on at a certain stage during the construction of a tableau. This emphasizes the view that the rôle of signs is on the meta level, namely to introduce certain non-classical properties of the model semantics into an otherwise classical deductive system. The ontological status that we give to signs here is quite similar to that of labels in Gabbay's Labelled Deductive Systems (Gabbay, 1991), though not quite so general.

Uniform Notation for tableau systems was introduced by Smullyan (1968) as an important technique that shortens and clarifies proofs and implementations. Although uniform notation does not work in the general setting of Chapter 4, one can define reasonable restrictions on the class of logics that will give us back uniform notation as a tool. This is done in Chapter 5, together with a justification, why we think the resulting class of logics (which we call *regular logics*) is important and what its scope within many-valued logics is. It will turn out that the concept of regular logics is closely connected with the existence of useful quantifier rules. Consequently, many-valued predicate logics are also dealt with in Chapter 5.

In Chapter 6 we show that the idea explicitly incorporating semantic information into a deduction system via sets of truth values is by no means restricted solely to the tableau method, nor does it make the tableau method incompatible with various improvements known from classical logic. We argue that sets of truth values as bits of meta information are a natural device to enhance the performance of most many-valued proof systems. We support our claim by sketching a possible way to enhance our tableau system, following recent suggestions of D'Agostino (1990) for classical logic. In Section 6.2 we show that a further generalization of our approach leads to a new translation from deduction to integer programming and the possibility of implementing many-valued theorem proving very efficiently, at least on the propositional level. Even resolution is amenable to sets-as-signs, as we are going to show in Section 6.3.4.

Finally, we will evaluate the proposed methods against the requirements catalogue set up in the Introduction.

3.1   Signed tableaux for classical logic

In this section we give a short account of semantic tableaux for classical logic. By a classical logic we mean a first-order logic whose associated propositional matrix has the truth value set **2**, with designated truth value 1. Common operators, such as conjunction, disjunction, material implication, negation, and so on, are denoted and interpreted as usual.

Tableaux systems come in two versions, namely signed and unsigned, from which we will always use the former for reasons which will soon become obvious.

In the classical case our set of signs (sometimes also called prefixes) will be $\{F, T\}$ where F, of course, corresponds to the truth value 0 and T to 1.

Definition 3.1. (Signed formula) *A **signed formula** is a string of the form $S\phi$, where ϕ is a (propositional or first-order) formula and S is either F or T. If **L** is the set of formulas in a logic, the set of **signed formulas** will be denoted by $\mathbf{L}^*$.*

Following Smullyan we divide the set of signed formulas into four classes: α for propositional formulas of conjunctive type, β for propositional formulas of disjunctive type, γ for quantified formulas of universal type, and, finally, δ for quantified formulas of existential type. Smullyan called this *unified notation*. It simplifies the presentation of rules and the proofs of various results considerably.

This classification is motivated by the *tableau expansion rules* which are associated with each signed formula. The rules characterize the assertion of a truth value (corresponding to its sign) to a formula by means of asserting truth values to its direct subformulas. For example, $T(\phi \wedge \psi)$ holds if and only if $T\phi$ and $T\psi$ hold simultaneously. The rules for the various combinations of signs and formula types are given schematically in Table 3.1. Premisses and conclusions are separated by a horizontal bar, while vertical bars in the conclusion denote different *extensions* which are to be thought of as disjunctions. The correspondence between formulas and rule types is shown in Table 3.2 on page 18. For convenience we treat negated formulas as α-formulas where $\alpha_1 = \alpha_2$.

Table 3.1 *Tableau rule schemata for different formula types*

α	β	γ	δ
$\dfrac{}{\begin{array}{c}\alpha_1 \\ \alpha_2\end{array}}$	$\dfrac{}{\beta_1 \mid \beta_2}$	$\dfrac{}{\gamma(t)}$	$\dfrac{}{\delta(c)}$
		where t is an arbitrary term.	where c is a new Skolem constant.

For our purposes it is sufficient to visualize a tableau proof as a finite labelled binary tree, whose node labels are signed formulas, constructed as follows.

Definition 3.2. (Classical tableaux) *Let $\mathbf{L}^*$ be a language of signed formulas. The set $\mathcal{T}(\mathbf{L}^*)$ of all **tableaux over** $\mathbf{L}^*$ is defined as the set of trees that can be constructed by finitely many applications of the following rules:*

(T1) A finite linear tree whose node labels are signed formulas is a tableau over $\mathbf{L}^*$.

(T2) If $\mathbf{T}$ is a tableau over $\mathbf{L}^*$ and ϕ is a node label from $\mathbf{T}$ then a new tableau $\mathbf{T}'$ is constructed by extending all branches of $\mathbf{T}$ that contain

Table 3.2   *Correspondence between rule types and formulas*

α	α_1	α_2
$\mathsf{T}\,(\phi \wedge \psi)$	$\mathsf{T}\,\phi$	$\mathsf{T}\,\psi$
$\mathsf{F}\,(\phi \vee \psi)$	$\mathsf{F}\,\phi$	$\mathsf{F}\,\psi$
$\mathsf{T}\,\neg\phi$	$\mathsf{F}\,\phi$	$\mathsf{F}\,\phi$
$\mathsf{F}\,\neg\phi$	$\mathsf{T}\,\phi$	$\mathsf{T}\,\phi$

β	β_1	β_2
$\mathsf{T}\,(\phi \vee \psi)$	$\mathsf{T}\,\phi$	$\mathsf{T}\,\psi$
$\mathsf{F}\,(\phi \wedge \psi)$	$\mathsf{F}\,\phi$	$\mathsf{F}\,\psi$

γ	$\gamma(t)$
$\mathsf{T}\,(\forall x)\phi(x)$	$\mathsf{T}\,\phi(t)$
$\mathsf{F}\,(\exists x)\phi(x)$	$\mathsf{F}\,\phi(t)$

δ	$\delta(c)$
$\mathsf{F}\,(\forall x)\phi(x)$	$\mathsf{F}\,\phi(c)$
$\mathsf{T}\,(\exists x)\phi(x)$	$\mathsf{T}\,\phi(c)$

ϕ by as many new linear subtrees as the rule* corresponding to ϕ has extensions, the nodes of the new subtrees being labelled with the formulas in the extensions.

If **T** is a tableau and Φ is the set in step (T1) above, then **T** will also be called a **tableau for** Φ.

Remark 3.3. *By the term 'new' in the δ-rule we mean that c is a symbol that does not occur in the tableau constructed so far. Therefore, the formulas which occur in a tableau proof tree are not merely from* **L**, *but actually from* **L**^{par}.

Definition 3.4. (Branch) *Let* **T** *be a tableau. A* **branch** $\mathbf{B}_T$ *of* **T** *is a maximal path in* **T**.

Usually, when no confusion arises, we omit the subscript from $\mathbf{B}_T$. Sometimes, when we speak of a branch **B**, we actually mean the set of node labels (signed formulas) on **B**; however, we still use the symbol **B**.

Definition 3.5. (Complementary signs and formulas)   *Two signs* S_1, S_2 *are* **complementary** *iff* $\mathsf{S}_1 \neq \mathsf{S}_2$. *Let* $\mathsf{S}_1\phi$, $\mathsf{S}_2\psi$ *be two formulas on a tableau branch* **B**. *They are called* **complementary formulas** *iff* S_1, S_2 *are complementary signs and* $\phi = \psi$.

Definition 3.6. (Closed and open branch) *A tableau branch is* **closed** *iff it contains a pair of complementary formulas. Otherwise it is called* **open**.

To prove that a formula ϕ is a tautology we begin with a tree whose single node is labelled with $\mathsf{F}\,\phi$, that is to say, we assume that ϕ is false in some model. A tableau proof represents a systematic search for such a model. Every tableau branch corresponds to a partial possible model in which the formulas on the branch are assigned the truth value corresponding to

---

*It is obtained by looking up the subformulas corresponding to ϕ in Table 3.2 and instantiating the matching rule schema in Table 3.1.

their sign. Therefore, a complementary pair of formulas, and consequently a closed branch, denotes an explicit contradiction, since in every model each formula has a unique truth value.

Definition 3.7. (Closed tableau, tableau provable) *A tableau* **T** *is* **closed** *iff all of its branches are closed. A* **L**-*formula* ϕ *is* **classically tableau provable,** *in symbols* $\vdash_c \phi$, *iff there exists a closed tableau for* $\{F \phi\}$.

Definition 3.8. (Complete branch, complete tableau)   *A tableau branch is called* **complete** *if it is either closed or no rule application to a formula on the branch produces a formula that was not already present. A tableau is complete iff each of its branches is.*

A tableau proof tree represents a proof of the negated root formula when all branches in the tree can be closed simultaneously; in other words, when every attempt to construct a model that makes the root formula false leads to a contradiction.

At this point a few remarks are in order:

Remark 3.9. (Closure of branches)

- *If 0-ary propositional connectives such as* **t** *(which evaluates constantly to 1) and* **f** *(which evaluates constantly to 0) are present (cf. the logic in Section 8.1.4.3), additional closure conditions for branches become necessary. The constant operators* **t**, **f** *are handled by letting branches also be closed when they contain one of the formulas* **T f, F t**.

- *Therefore, the following alternative definition of branch closure is possible which will prove useful in the following chapter.*

  Alternative definition of branch closure
  Let **Contr**$_c$ = $\{\{T \phi, F \phi\}\mid \phi \in$ **L**$\} \cup \{\{T \mathbf{f}\}, \{F \mathbf{t}\}\}$ *be the* **contradiction set** *for classical tableaux. Then a tableau branch* **B** *is* **closed** *iff* $2^{\mathbf{B}} \cap$ **Contr**$_c \neq \emptyset$.

- *From now on we will use this definition. In all tableau systems for the various logics that we will consider, what will change besides the formula syntax are the tableau expansion rules and the choice of the contradiction set.*

- *It is sufficient to consider complementary pairs of atomic formulas in the definition of branch closure.*

Theorem 3.10. (Soundness, completeness) *Let* $\mathcal{L}$ *be a classical first-order logic and let* ϕ *be any* **L**-*sentence. Then there is a closed classical tableau for* $\{F \phi\}$ *iff* ϕ *is a first-order classical tautology. In symbols,*

$$\models_{\mathcal{L}} \phi \text{ iff } \vdash_c \phi.$$

Proof. See, for example, Fitting (1990b).

Remark 3.11. (Deletion of used formulas) *It is sufficient for completeness to apply α-,β- and δ-rules only once to every formula in each branch. Consequently, formulas of these types may be deleted locally to the current branch after a rule has been applied to them. Note, however, that in general γ-formulas must be used repeatedly and hence may not be removed.*

Remark 3.12. (Systematic tableaux and fairness) *Tableau construction for a set of formulas Φ as described above is a highly non-deterministic procedure. We did not specify, for example, in which order the tableau rules should be applied to the formulas on a branch, in which order newly generated branches should be processed, or what terms in which order should be 'guessed' by the γ-rules. Somewhere on the way to an actual implementation, however, these questions have to be addressed, since any real program on a real machine behaves deterministically. Our completeness result, on the other hand, does not exhibit anything of the order in which the tableau is built up.*

If one wishes to extend the completeness result towards a concrete implementation the notion of systematic tableaux (Smullyan, 1968) is usually introduced. The main issues to be considered are the γ-rule applications. If the γ-rule is repeatedly applied to the same formula every term in the language must finally occur in the conclusion. Moreover, the order of rule applications to γ-formulas are to be 'fair'. Basically, this means that a tableau construction is carried out in such a way that for every γ-formula the corresponding rule is applied arbitrarily often. Depending on the rules used, fairness conditions can become quite complex, in particular when other optimizations, such as indexing of formulas, are implemented (Hähnle et al., 1992).

Since fairness issues are not dependent on the number of truth values in a logic, we will not discuss them in the present work. We merely inform the reader that all our completeness results could be strengthened with respect to systematic tableaux and we refer to Smullyan (1968), Fitting (1990b), Hähnle (1992) for a deeper discussion of fairness.

Remark 3.13. (Ground vs free variable tableaux) *An important optimization in recent tableau-based theorem provers deals with the γ-rules. Instead of guessing an arbitrary ground term, as is done in the γ-rule we are using here, one marks the quantified variable in the conclusion as free and it is instantiated only later via unification with a term that is actually needed for a branch closure. Of course, then something also has to be done with the δ-rules in order to preserve soundness. For obvious reasons we speak of a free variable tableau system when rules of this kind are used and of a ground tableau system when the present rules are used. We will take up this issue again in Section 6.3.1.2.*

Remark 3.14. (Strong soundness and completeness)   *In classical logic the usual strong soundness and completeness results can easily be proved by observing that for all first-order sentences $\phi_1, \ldots, \phi_n, \psi$ and classical logics $\mathcal{L}$*

$$\{\phi_1, \ldots, \phi_n\} \vDash_{\mathcal{L}} \phi \ \textit{iff} \ \vDash_{\mathcal{L}} (\phi_1 \wedge \ldots \wedge \phi_n) \supset \phi$$

holds. This fact, which is a kind of deduction theorem, does not hold in most many-valued logics, and worse, a consequence relation is not necessarily characterizable by finite matrices at all (cf. Section 8.1.3). In the following, we will not, therefore, be concerned with consequences, but only with tautologies.

We conclude this section with two small examples whose purpose is merely to introduce our notation for tableau proof trees.

Example 3.15. *Using the tree drawn on the left we prove the first-order tautology $\vdash_c (\exists x)(\forall y)r(x, y) \supset (\forall y)(\exists x)r(x, y)$, while the tree on the right proves $\vdash_c p \supset (q \supset p)$. Formulas marked with an asterisk may be removed during the construction.*

```
 * (1) [−]   F (∃x)(∀y)r(x,y) ⊃ (∀y)(∃x)r(x,y)        * (1) [−]   F p ⊃ (q ⊃ p)
                          |                                              |
 * (2) [1]   T (∃x)(∀y)r(x,y)                            (2) [1]   T p
                          |                                              |
 * (3) [1]   F (∀y)(∃x)r(x,y)                          * (3) [1]   F q ⊃ p
                          |                                              |
   (4) [2]   T (∀y)r(c,y)                                (4) [3]   T q
                          |                                              |
   (5) [3]   F (∃x)r(x,d)                                (5) [3]   F p
                          |                                           closed by (2, 5)
   (6) [4]   T r(c,d)
                          |
   (7) [5]   F r(c,d)

           closed by (6, 7)
```

The formulas on the trees are numbered in the order of their appearance, starting with (1). These numbers are enclosed in parentheses. The numbers in square brackets indicate the number of the parent formula. Beneath each closed branch the numbers of the formulas which led to the closure are given.

On the left side, in (4) a new parameter c has been introduced via a δ-rule application to (2), while (6) was inferred from (4) by a γ-rule application, whereby y was instantiated by d. Similarly, (7) was inferred from

(5) *and* (5) *in turn from* (3). *The first rule applied to the tree was an α-rule to* (1), *corresponding to* F *and implication, which produced formulas* (2) *and* (3).

The example on the right should be obvious enough. It is part of most Hilbert-style axiomatizations of classical logic and it will be used several times throughout the book, since it is also valid in various many-valued logics.

For some more sophisticated examples of classical proof trees, see (Smullyan, 1968).

3.2   On the rôle of signs in analytic tableaux

The notion of a *sign* or *prefix* is central to our approach to many-valued theorem proving; unlike classical logic, where the choice between signed and unsigned tableau systems is merely a matter of taste (since they are completely dual), in non-classical logics signs are crucial. To stress this point we assume from now on that the set of signs is specified together with the language. Consequently, from now on, we will use the following definition instead of Definition 2.8, at least when tableaux are involved.

Definition 3.16. (Propositional logic) *Let* **L** *be a propositional language,* **A** *a propositional matrix for it, and* **S** *a finite set of signs with* $\mathbf{L} \cap \mathbf{S} = \emptyset$. *Then the triple* $\mathcal{L} = (\mathbf{L}, \mathbf{A}, \mathbf{S})$ *is called an n*-**valued propositional logic** *with designated truth values D.*

Signed formulas are defined as above, but the signs are elements from **S** for which we sometimes also write $\mathbf{S}_{\mathcal{L}}$ to emphasize that the choice of signs is part of the definition of a logic. We feel this is justified, since changing the set of signs not only leads to a different proof system, but also forces one to express queries differently, and thus has an influence on the form of completeness theorems, as we will see.

We said that changing the set of signs will result in a different proof system, just as does changing the set of connectives, however, the analogy carries further: while a certain set of connectives has to be present for a logic to be functionally complete, a certain set of signs also has to be present for the existence of sound tableau systems.

We think it is an important step in proof procedures for many-valued logics (and not only for these) to separate the link structure, which is purely classical since it amounts to a semantic And/Or-graph over signed formulas, from the nonclassical part, which can be described as bookkeeping or resource management; and signs are a natural device to achieve this. In a way the sign language then has to reflect the kind of deviation from classical logic. In our case, truth value sets as signs reflect many-valued matrices; in other logics they may reflect temporal relationships or resource constraints. A similar philosophy is also behind Gabbay's work on Labelled Deductive Systems (Gabbay, 1991).

Seen in this way, signs are not restricted to use within a tableau framework, but can be used advantageously in tableaux as well as in other proof procedures. We will illustrate this latter point for Binary Decision Diagrams and path dissolution, KE systems and resolution in Chapter 6.

3.3   Multiple-valued extension of tableau systems

In this section we sketch a method* of extending tableaux to handle any finitely-valued first-order logic. The method is due to Surma, who presented it at the International Symposium on Multiple-Valued Logic in 1974 (Surma, 1984). In the same year Suchoń presented a tableau system for the special case of n-valued Łukasiewicz logics which had the advantage of yielding much shorter proofs than one would normally obtain using Surma's approach. We delay the discussion of Suchoń's system until Section 7.2.3, where we will describe precisely the relationship between both approaches using the tools that will have been developed by then.

Some years later, Carnielli (1987) filled the gaps in Surma's somewhat sketchy presentation and extended the treatment to distribution quantifiers (Mostowski, 1957) which include the standard ones that we consider.†

Assume for the moment that we are working in a three-valued logic. Then, obviously, stating that a formula ϕ is not true is not the same thing as stating that it is false, or, more precisely, stating that ϕ does not have the truth value 1 is not equivalent to stating that it has the truth value 0. Yet another formulation of this fact, with respect to signed tableaux, is that *not* T ϕ is not the same as F ϕ. But being able to express this fact, that *not* T ϕ holds, is crucial for the tableau method to work, since this is what we must put in the initial tableau if the tautological status of a formula ϕ is to be established.

The solution is to introduce signs other than T and F , namely as many as there are truth values in a logic. Let us fix

$$\mathbf{S}_{\mathcal{L}} = \left\{ 0, \frac{1}{n-1}, \dots, \frac{n-2}{n-1}, 1 \right\}$$

as the set of signs for an n-valued logic in this section. Each sign corresponds to a truth value in an obvious way. We use the same symbols for signs and truth values. For better readability the former are usually printed in sans serif typeface and this convention is understood as an implicit conversion function. We are now able to express invalidity of a formula in the following way:

---

*In the next chapter we will see that it can be regarded as a special case of the technique we are developing. Since we give the latter a thorough treatment we can afford to be sloppy in the present section.

†The history of many-valued proof systems based on tableaux and Gentzen calculi is actually a bit longer. See Section 8.2.2 for some more references.

$$not\ 1\ \phi \text{ iff } (0\ \phi \text{ or } \tfrac{1}{2}\ \phi) \qquad\qquad (3.1)$$

So far, so good—but how does one compute the tableau rules corresponding to a certain sign and connective in this new setting? Assume that we wanted to compute the rule corresponding to $\tfrac{1}{2}$ and disjunction for $n = 3$. If we take a look at the truth table (Table 2.1) we see that there are three entries equal to $\tfrac{1}{2}$. By covering these entries we can extract the necessary and sufficient conditions on the direct subformulas of a formula of the form $\phi \vee \psi$ that characterize the assertion $\tfrac{1}{2}\ (\phi \vee \psi)$. More precisely, we have that

$$\tfrac{1}{2}\ (\phi \vee \psi) \quad \text{iff} \quad (\tfrac{1}{2}\ \phi \text{ and } 0\ \psi) \text{ or } (\tfrac{1}{2}\ \phi \text{ and } \tfrac{1}{2}\ \psi) \text{ or } (0\ \phi \text{ and } \tfrac{1}{2}\ \psi).$$

Transforming this into a tableau rule, we obtain

$$\frac{\tfrac{1}{2}\ (\phi \vee \psi)}{\begin{array}{c|c|c} \tfrac{1}{2}\ \phi & \tfrac{1}{2}\ \phi & 0\ \phi \\ 0\ \psi & \tfrac{1}{2}\ \psi & \tfrac{1}{2}\ \psi \end{array}}$$

Recall that formulas on the same tableau branch are to be thought of as conjunctively connected while formulas on different branches are disjunctively connected, hence the one-to-one correspondence between conclusions of tableau rules and disjunctive normal forms of their premises.

Using the same method one can compute rules for all combinations of signs and connectives, provided the truth tables are known and the sign *does* occur in the truth table of the connective. If it does not, as, for example, the sign $\tfrac{1}{2}$ in the truth table for weak negation (Table 2.2), the current branch can be closed immediately since an assertion of this kind can never be satisfied. We call signed formulas of this kind **self-contradictory**. A convenient method of handling these cases is to include them in the contradiction set of a tableau system. We illustrate the point by giving the contradiction set for the logic $\mathcal{L}_M^n$ (see Definition 2.28).

$$\textbf{Contr}_{\mathcal{L}_M^n} = \{\{S_1\phi, S_2\phi\}|\ \phi \in \mathbf{L}_M^n, S_1 \neq S_2\}$$
$$\cup\{k\ J_i(\phi)|\ 0 < k < 1, \phi \in \mathbf{L}_M^n,\ i, k \in N\}$$

Thus every branch that contains contradictory formulas, either complementary pairs or self-contradictory formulas, can be closed.

One can note at once some important properties of many-valued tableaux:

1. The rules do not in general fall into either the α- or the β-schema.
2. The number of extensions generated by a rule for a formula $S\phi$ can be equal to the number of entries corresponding to S in the truth table of the top connective of ϕ. For n truth values and k-ary connectives the worst case is a branching factor of $n^k - n$ with $k(n^k - n)$ formulas

in the conclusion. Since every entry in the truth table has to be analysed in exactly one of the rules corresponding to the connective, the number of extensions summarized over all rules corresponding to a connective is, with the exception of rare cases, when simplifications are possible, equal to n^k.

3. For connectives with an arity not greater than 2 the rules are uniquely determined up to the ordering of the extensions in the conclusion and the formulas within each extension.*

4. Using (3.1) we can stipulate that a formula is not satisfiable. The price, however, is that the construction of *two* tableaux is required. In general, as many tableaux as there are non-designated truth values are required.

Example 3.17. *In Figures 3.1 and 3.2 the proof trees corresponding to the proof of the $\mathcal{L}_M^3$-tautology (see Table 2.1) $\neg p \supset (\sim p \land \neg p)$ are shown. Note that we need two proof trees in order to refute the two non-designated truth values 0 and $\frac{1}{2}$. Formulas in Figure 3.1 correspond to formulas with the same numbers in Figure 3.2.*

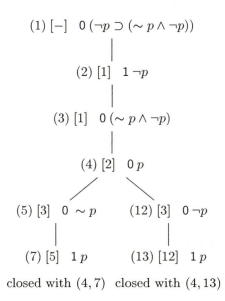

(1) [−] 0 (¬p ⊃ (∼ p ∧ ¬p))
|
(2) [1] 1 ¬p
|
(3) [1] 0 (∼ p ∧ ¬p)
|
(4) [2] 0 p

(5) [3] 0 ∼ p   (12) [3] 0 ¬p
| |
(7) [5] 1 p   (13) [12] 1 p

closed with (4, 7)   closed with (4, 13)

FIG. 3.1. Tableau proof of $\neg p \supset (\sim p \land \neg p)$ using Surma and Carnielli's method: refutation of truth value 0.

---

*For connectives with higher arity this is generally not the case even in classical logic. See Section 4.5 for an example.

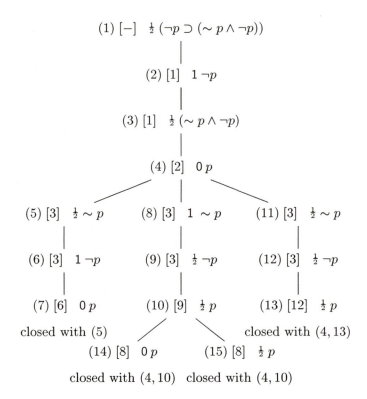

FIG. 3.2. Tableau proof of $\neg p \supset (\sim p \wedge \neg p)$ using Surma and Carnielli's method: refutation of truth value $\frac{1}{2}$.

Carnielli (1987; 1991) admits a very general class of quantifiers of which $\forall$ and $\exists$ are special cases. Assume for the moment a first-order language **L** containing quantifiers $\mathbf{Q} = \{Q_1, \ldots, Q_r\}$ and thus formulas $(Q_i x)\phi$.

Definition 3.18. (Distribution of a formula) *Let $\phi \in \mathbf{L}$ and β be a variable assignment in a structure $\mathbf{M}$ for $\mathcal{L}$. Then the **distribution** of ϕ with respect to $\mathbf{M}$ and β is defined as*

$$\mathbf{Distr}_{\mathbf{M},\beta}(\phi) = \{v_{\beta_x^u}(\phi) | \ u \in U\}$$

where U is the domain of $\mathbf{M}$.

Thus the distribution of ϕ yields all truth values that ϕ can take on under $\mathbf{M}$ and β while x is running through all possible values. Obviously, $\mathbf{Distr}_{\mathbf{M},\beta}(\phi) \subseteq N$ always holds. Now it is possible to define the semantics of a quantifier with a mapping from distributions to truth values, in other words, each $d_Q : 2^N \to N$ defines a quantifier Q by the following convention, which replaces clauses (3) and (4) in Definition 2.19:

$$v_\beta((Q_i x)\phi) = d_{Q_i}(\mathbf{Distr_{M,\beta}}(\phi))$$

The resulting quantifiers are called, appropriately enough, **distribution quantifiers**. They were first introduced by Mostowski (1957). The familiar quantifiers $\forall$ and $\exists$ may be defined using the mappings

$$d_\forall : X \mapsto \min(X)$$

$$d_\exists : X \mapsto \max(X)$$

where min and max are interpreted naturally on $X \subseteq N$. If we define d_Q^{-1} as

$$d_Q^{-1}(i) = \{X \mid \emptyset \neq X \subseteq N, d_Q(X) = i\}$$

tableau rules for a sign i and a closed formula $(Q\,x)\phi$ are easily obtained by analysing the possibilities in $d_Q^{-1}(i)$ on a branch-by-branch basis. Informally, a set of truth values $\{i_1, \ldots, i_k\} \in d_Q^{-1}(i)$ says that

1. $v(\phi\{x \leftarrow t_1\}) = i_1, \ldots, v(\phi\{x \leftarrow t_k\}) = i_k$ holds for certain terms $\{t_1, \ldots, t_k\}$ and
2. for any term t: $v(\phi\{x \leftarrow t\}) \in \{i_1, \ldots, i_k\}$.

We observe that if $k = 1$ then for all terms t we have $v(\phi\{x \leftarrow t\}) = i_1$.

A tableau rule that reflects the first condition for a sign i and a formula $(Q\,x)\phi$ has $|d_Q^{-1}(i)|$ many extensions, one for each set $\{i_1, \ldots, i_k\} \in d_Q^{-1}(i)$. If $k > 1$ an extension contains the signed formulas $i_1\ \phi\{x \leftarrow c_1\}, \ldots, i_k\ \phi\{x \leftarrow c_k\}$, where $\{c_1, \ldots, c_k\}$ are new parameters. If $k = 1$ then an extension contains the single formula $i_1\ \phi\{x \leftarrow t\}$, where t is an arbitrary term.

The contradiction set must be extended to include the following signed formulas:

$$\{i\ (Q\ x)\phi \mid d_Q^{-1}(i) = \emptyset, \phi \in \mathbf{L}, Q \in \mathbf{Q}\}$$

Unfortunately, this definition from Carnielli (1987) as it stands can lead to incomplete rules, since the second condition above is not taken into account. An example that shows this as well as a possible correction is given in Carnielli (1991).

Example 3.19. (Carnielli, 1991) *Let $\mathcal{L}$ be a three-valued logic containing two quantifiers Q, R which are defined as follows:*

$$d_Q(X) = \begin{cases} 1 & X = \{0, \frac{1}{2}\} \\ 0 & otherwise \end{cases} \qquad d_R(X) = \begin{cases} 1 & X = \{0, 1\} \\ 0 & otherwise \end{cases}$$

*If we use the construction of tableau rules just given we arrive at the
following rules for the prefix 1 :*

$$\frac{1\,(Q\,x)\phi(x)}{\begin{array}{l}0\,\phi(c_1)\\ \tfrac{1}{2}\,\phi(c_2)\end{array}} \qquad\qquad \frac{1\,(R\,x)\phi(x)}{\begin{array}{l}0\,\phi(c_3)\\ 1\,\phi(c_4)\end{array}}$$

where the c_i are new parameters.

*Now consider the set of signed formulas $\Phi = \{1\,(Q\,x)p(x), 1\,(R\,x)p(x)\}$
for an arbitrary predicate symbol p. Despite the fact that Φ is unsatisfiable,
using these rules we obtain the tableau shown in Figure 3.3 which cannot
be closed.*

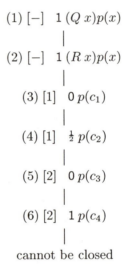

$$(1)\;[-]\quad 1\,(Q\,x)p(x)$$
$$|$$
$$(2)\;[-]\quad 1\,(R\,x)p(x)$$
$$|$$
$$(3)\;[1]\quad 0\,p(c_1)$$
$$|$$
$$(4)\;[1]\quad \tfrac{1}{2}\,p(c_2)$$
$$|$$
$$(5)\;[2]\quad 0\,p(c_3)$$
$$|$$
$$(6)\;[2]\quad 1\,p(c_4)$$
$$|$$

cannot be closed

FIG. 3.3. Attempted tableau refutation of Φ with incomplete quantifier
rules.

The incomplete rules are fixed in Carnielli (1991) by introducing additional rules with an empty premise (i.e. they can be applied at any time)
that list exhaustively the possible truth values of formulas with parameters
introduced at earlier stages. In the proof tree of Figure 3.3, for example,
one could apply (among many others) the following rule:

$$\frac{}{0\,p(c_2)\ \mid\ 1\,p(c_2)}$$

in order to close the tree. We argue that this solution introduces more in-
determinacy than necessary, causing excessive branching and complicating

the generated proofs. In Section 5.4 we give a more satisfactory solution which avoids most of the drawbacks.

To further illustrate how complex the computation of first-order tableau rules can be using Carnielli's approach we consider the standard quantifiers, say $0 \, (\forall x)\phi(x)$, in three-valued logic. From

$$d_\forall^{-1}(0) = \{\{0\}, \left\{0, \frac{1}{2}\right\}, \left\{0, \frac{1}{2}, 1\right\}, \{0, 1\}\}$$

the following tableau rule is computed:

$$\frac{0 \, (\forall x)\phi(x)}{}$$

$0 \, \phi(t)$	$0 \, \phi(c_1)$	$0 \, \phi(c_3)$	$0 \, \phi(c_6)$
	$\frac{1}{2} \, \phi(c_2)$	$\frac{1}{2} \, \phi(c_4)$	
		$1 \, \phi(c_5)$	$1 \, \phi(c_7)$

Where t is any term and c_i are new parameters.

Since any distribution (i.e. non-empty subset of N) of a quantifier Q has to be analysed in exactly one of the rules corresponding to $0 \, (Q \, x)\phi, \ldots, 1 \, (Q \, x)\phi$, the sum of the extensions in all rules for one quantifier is always equal to $2^n - 1$ if the rules are computed by the method just sketched. Sometimes rules may be simplified, for example the rule for $0 \, (\forall x)\phi(x)$ stated above can be reduced to

$$\frac{0 \, (\forall x)\phi(x)}{0 \, \phi(c)} \qquad \text{where } c \text{ is a new parameter.}$$

It is a non-trivial task, however, to detect and validate possible simplifications, and Carnielli does not give a hint as to how to achieve them. Moreover, we did not count the branches generated by the additional rules that enumerate parameters.

3.4 Discussion

We have seen that using Surma and Carnielli's method it is possible to give tableau proof systems for every finitely-valued first-order logic including distribution quantifiers.* We have, however, also encountered obstacles which make the actual use of the method in a theorem prover highly problematic.

- The number of proof trees that are necessary to validate a formula is $|N - D|$ and thus usually $\mathcal{O}(n)$, since D is small.
- Given that signs and connectives are approximately equally distributed in the proof trees, the average number of new branches that are generated in a rule application is $\frac{n^k}{n} = n^{k-1}$ and the average number

---

*In Caferra and Zabel (1990) an implementation of a theorem prover based on Surma and Carnielli's approach for propositional logic is described. See also Section 7.2.4 of the present work.

of new formulas is $\frac{kn^k}{n} = kn^{k-1}$ for k-ary propositional connectives, while the figures for quantifiers are $\frac{2^n-1}{n}$ for the number of extensions and

$$\frac{\sum_{j=0}^n j\binom{n}{j}}{n} = \frac{\sum_{j=0}^n n\binom{n-1}{j-1}}{n} = \sum_{j=1}^n \binom{n-1}{j-1} = \sum_{i=0}^{n-1} \binom{n-1}{i} = 2^{n-1}$$

for the number of new formulas. The number of branches generated by the parameter rules depends on the number of parameters present in the tableau, but can be exponential wrt this number.[*]

- A closer inspection of the proof trees in Figures 3.1 and 3.2 reveals that all unsigned formulas in the tree shown in Figure 3.1 also occur in the tree shown in Figure 3.2 and at the same position: the first tree is isomorphic to a partial tree (that is, a tree that results when certain nodes are removed) of the second tree. Inspection of other examples shows that there is always a very high degree of redundancy in the trees corresponding to the various non-designated truth values.

From these considerations several questions naturally arise:

- Is there an economic representation of the many-valued search space which is less redundant?
- Is there a method decreasing the excessive branching factor of the rules, in particular for the first-order case?
- The quantifier rules are branching most and are complicated to denote, while at the same time it does not really seem to be useful to have the whole class of distribution quantifiers available. Is there a more restricted class of quantifiers that has more elegant and less branching rules instead?
- It is annoying that the many-valued rules do not obey Smullyan's classification schema into rules of type $\alpha, \beta, \ldots$. Is there a class of many-valued logics for which a uniform system does exist?

In the following two chapters we will develop methods that allow us to answer all of these questions in the affirmative. In Section 6.2 we will also give a tentative answer to the following related question:

Some many-valued logics seem to be more complex than others[†] and this should be mirrored in the corresponding proof systems, but in Surma and Carnielli's approach the proof complexity grows uniformly with the number of truth values (this is another indicator that they are far from optimal). Are there criteria for

---

[*]This analysis does not account for possible simplifications of the rules.

[†]For example, the finitely-valued Łukasiewicz logics (Section 2.3.2), which are based on the $\supset_L$ connective seem at first sight to be more complicated than the logics from the class $\mathcal{L}_M^n$ (Definition 2.28) which are based on the simple generalized disjunction connective. It will become apparent that this is not the case.

estimating the proof size complexity of a many-valued logic which are based on semantical properties and which go beyond the specialized setting of analytic tableaux?

4

A NEW TECHNIQUE: TRUTH VALUE SETS AS SIGNS

Glückseliger Aspekt! So stellt sich endlich
Die große Drei verhängnisvoll zusammen . . .
— Friedrich Schiller, Wallensteins Tod

As before stated we defer the discussion of first-order logics until the next chapter and deal only with propositional logics in the present one. Moreover, we assume that the languages under consideration contain no 0-place connectives. These are possible to accommodate, although they make the definitions unnecessarily complicated. We sketch what must be changed at the end of Section 4.3.

4.1 Sets as signs

One approach to decrease redundancy in the tableau systems introduced in the previous chapter would be to perform the steps that are identical in all or in some of the proof trees at the same time (possibly using structure sharing); in other words, to search for the refutation of all non-designated truth values in parallel. Now, as always, when one is making algorithms and data structures trickier, this leads to a fairly complex proof procedure involving much bookkeeping and hence a cryptic completeness proof. Moreover, the results achieved in this way are unlikely to be applicable outside the context of analytic tableaux. A far more satisfying solution can be achieved on a logical level.

Consider, for example, the signed $\mathbf{L}^3_{\mathrm{M}+}$-formula* $1 \sim \phi$. Application of the corresponding tableau rule computed according to Surma and Carnielli yields two new branches, each containing one of the formulas $0\,\phi$ and $\frac{1}{2}\,\phi$. Encountering such a formula during a proof, however, does not give rise to any proof theoretical reason to split the proof tree at once into the two cases determined by the extensions of the rule. If we were able to express the more complex assertion

ϕ has either truth value 0 or truth value $\frac{1}{2}$

with a single signed formula we could avoid splitting. In other words, we *delay the final decision of which model to assume for ϕ while retaining all necessary information*. Hence, our idea is to increase the expressivity of the

---

*For the logic $\mathcal{L}^3_{\mathrm{M}+}$ see Definition 2.29.

sign language in order to be able to state more complex conditions like the one just shown. Perhaps the most natural thing to do is to admit *subsets of the set of truth values* as signs.

Definition 4.1. (Base set of signs) *The* **base set of signs** $\bar{\mathsf{S}}$ *is defined as*

$$\bar{\mathsf{S}} = \{\{\mathsf{k}_1, \ldots, \mathsf{k}_m\}| \ \{k_1, \ldots, k_m\} \subseteq N\} = 2^N$$

For the moment assume that the set of signs $\mathbf{S}_{\mathcal{L}}$ in a logic $\mathcal{L}$ always obeys

$$\{\{i\}|\, i \in N\} \subseteq \mathbf{S}_{\mathcal{L}} \subseteq \bar{\mathbf{S}} \tag{4.1}$$

We need the left-hand part of (4.1), because, otherwise, unsound rules can be stated. In Section 4.2 we give a weaker condition that is sufficient for soundness.

Example 4.2. *Let us fix the set of signs for the logic* $\mathcal{L}^3_{\mathrm{M+}}$ *as*

$$\mathbf{S}_{\mathcal{L}^3_{\mathrm{M+}}} = \{\{0\}, \{\tfrac{1}{2}\}, \{1\}, \{0, \tfrac{1}{2}\}, \{\tfrac{1}{2}, 1\}\}$$

We can express the assertion that ϕ *has either truth value 0 or truth value* $\tfrac{1}{2}$ *by the signed formula* $\{0, \tfrac{1}{2}\}\ \phi$. *An equivalent formulation would be to say that* ϕ *cannot take on truth value 1, hence we need to build only one tableau proof tree for each proof.*

We turn now to the question of how the tableau rules for generalized signs are computed. We can view the process of finding the extensions to a given sign S and connective F in a similar way as before (cf. Section 3.3), namely as finding of cover for all entries in the truth table of F that are members of S. The difference from the former approach is that, in general, extensions in which formulas with generalized signs do occur cover more than one entry.

The next definitions formally describe this process. Remember that syntax and semantics are defined in terms of abstract algebras and valuations are homomorphisms between them.

Definition 4.3. (Algebra of signs) *Let* $\mathcal{L} = (\mathbf{L}, \mathbf{A}, \mathbf{S})$ *be any propositional logic with matrix* $\mathbf{A} = (N, f_1, \ldots, f_r)$. *Then we define the* **algebra of signs** $\mathbf{A}_{\mathrm{S}} = (\mathbf{S}, f'_1, \ldots, f'_r)$ *as an abstract algebra similar to* $\mathbf{A}$ *whose operations* f'_i *are defined as*

$$f_i'(\mathsf{S}_1,\ldots,\mathsf{S}_m) = \bigcup\{f_i(j_1,\ldots,j_m)|\ j_k \in \mathsf{S}_k, 1 \le k \le m\}$$

The algebra $\mathbf{A}_\mathsf{S}$ defines the semantics of $\mathcal{L}$ in terms of truth value sets corresponding to the members of $\mathbf{S}$. In the following definition we use the convention $f^{-1}(\mathsf{S}) = \{(j_1,\ldots,j_m)|\ f(j_1,\ldots,j_m) \in S\}$, where S is the set of truth values corresponding to S.

Definition 4.4. (Tableau rule) *Let* $\mathcal{L} = (\mathbf{L},\mathbf{A},\mathbf{S})$ *be a many-valued logic,* $\phi = F(\phi_1,\ldots,\phi_m)$ *a* $\mathcal{L}$*-formula, and let* f *and* f' *be the interpretations of* F *in* $\mathbf{A}$, *respectively in* $\mathbf{A}_\mathsf{S}$.

A ($\mathcal{L}$-)**tableau rule** *is a function* $\pi_{\mathsf{S},F}$ *which assigns to the signed formula* $\mathsf{S}\phi \in \mathbf{L}^*$ *a tree with root* $\mathsf{S}F(\phi_1,\ldots,\phi_m)$ *called* **premise** *and linear subtrees (denoted by* $\circ\cdots\circ$*)* $\mathsf{S}_1\,\phi_{i_1} \circ \cdots \circ \mathsf{S}_t\,\phi_{i_t}$ *such that* $\mathsf{S}_1,\ldots,\mathsf{S}_t \in \mathbf{S}$, *where* $1 \le t \le m$ *and the condition* $H_\mathsf{S}(F;(\mathsf{S}_1,i_1),\ldots,(\mathsf{S}_t,i_t))$, *which is defined below, holds for each of these subtrees.*

The linear subtrees are called **extensions**.* *The collection of extensions is called a* **conclusion** *of a tableau rule if*

(T0a) *for any* $(j_1,\ldots,j_m) \in f^{-1}(\mathsf{S})$ *there is an extension* $\mathsf{S}_1\phi_{i_1} \circ \ldots \circ \mathsf{S}_t\phi_{i_t}$ *with* $j_{i_k} \in S_k$ *for all* $1 \le k \le t$, *and*

(T0b) *there is no set of extensions with fewer elements satisfying (T0a).*

$H_\mathsf{S}(F;(\mathsf{S}_1,i_1),\ldots,(\mathsf{S}_t,i_t))$ *holds iff there exists a homomorphism* $h : \mathbf{L} \to \mathbf{A}_\mathsf{S}$, *satisfying (T1)–(T4).*

(T1) $h(\phi_{i_k}) = \mathsf{S}_k$ *for* $1 \le k \le t$.

(T2) $f'(\mathsf{S}_1',\ldots,\mathsf{S}_m') \subseteq S$ *whenever* $\mathsf{S}_{i_k}' = \mathsf{S}_k$ *for all* $1 \le k \le t$ *and the other* S_j' *are arbitrary.*

(T3) *For no* $1 \le k \le t$ *is there a* S_k' *with* $\mathsf{S}_k' \not\supsetneq \mathsf{S}_k$ *that satisfies (T1) and (T2).*

(T4) *There is no* t' *with* $t' < t$ *that satisfies (T1) and (T2).*

If no such homomorphism exists, no rule for the formula ϕ *and the sign* S *is defined.*

Since this rule is fairly abstract, we feel that some explanations are in order:

ad (T0) The first part specifies soundness of rules with respect to $\mathbf{A}_\mathsf{S}$. Each truth table entry which is a member of S must be covered by some extension. The second part minimizes the number of extensions. This is already necessary in the two-valued case. Three versions of a sound and complete rule for $\{1\}$ ($\phi \vee \psi$) in

---

*Extensions are treated as sets. Thus of all subtrees that differ only in the ordering of their signed formulas only one appears as an extension of the rule.

classical logic are shown in Figure 4.1. (T0b) ensures that the one on the right is never selected.

ad (T1)   Since h is a homomorphism it respects the semantics of $\mathcal{L}$. (T1) propagates this property to the rule via h.

ad (T2)   This guarantees completeness with respect to $\mathbf{A_S}$.

ad (T3)   Makes the S_k maximal and favours, for example, the rule for three-valued disjunction and $\{\frac{1}{2}\}$ on the left in Figure 4.2 before the rule on the right.

ad (T4)   Minimizes the number of formulas in extensions. Once again, this is already necessary in the two-valued case. The rule in the middle of Figure 4.1 is favoured before the rule on the left.

$$
\frac{\{1\}\,(\phi \vee \psi)}{\begin{array}{c|c} \{1\}\,\phi & \{0\}\,\phi \\ & \{1\}\,\psi \end{array}}
\qquad
\frac{\{1\}\,(\phi \vee \psi)}{\{1\}\,\phi \mid \{1\}\,\psi}
\qquad
\frac{\{1\}\,(\phi \vee \psi)}{\begin{array}{c|c|c} \{1\}\,\phi & \{1\}\,\phi & \{0\}\,\phi \\ \{0\}\,\psi & \{1\}\,\psi & \{1\}\,\psi \end{array}}
$$

FIG. 4.1. Various tableau rules for classical disjunction and sign $\{1\}$.

$$
\frac{\{\tfrac{1}{2}\}\,(\phi \vee \psi)}{\begin{array}{c|c} \{0,\tfrac{1}{2}\}\,\phi & \{\tfrac{1}{2}\}\,\phi \\ \{\tfrac{1}{2}\}\,\psi & \{0,\tfrac{1}{2}\}\,\psi \end{array}}
\qquad
\frac{\{\tfrac{1}{2}\}\,(\phi \vee \psi)}{\begin{array}{c|c} \{0,\tfrac{1}{2}\}\,\phi & \{\tfrac{1}{2}\}\,\phi \\ \{\tfrac{1}{2}\}\,\psi & \{0\}\,\psi \end{array}}
$$

FIG. 4.2. Two tableau rules for three-valued disjunction and sign $\{\tfrac{1}{2}\}$.

From the multitude of sound and complete rules that are possible for each signed formula the above definition selects one with the help of the minimizing and maximizing conditions (T0b), (T3), and (T4), which is still not unique up to the order of formulas in the extensions and the order of extensions in the conclusion. See Section 4.5 for an example of a connective that gives rise to non-unique rules. There we will also prove the following proposition which states that conditions (T0a), (T1), and (T2) alone are sufficient to guarantee the existence of sound and complete rules.

Proposition 4.5. *If there are homomorphisms* $g_1, \ldots, g_r$ *satisfying conditions (T0a), (T1), and (T2) in Definition 4.4, then there are also homomorphisms* $h_1, \ldots, h_q$ *with* $q \leq r$ *satisfying (T0)–(T4).*

It is possible, of course, to discuss conditions (T0b), (T3), and (T4), because they are not the only ones one could think of. In particular one could argue that (T3) should be reversed so that the rule on the right in Figure 4.2 is selected instead of the rule on the left, since the former gives more information about the formula ψ in the right extension than the latter.

A similar argument can be applied to (T4) and the two leftmost rules in Figure 4.1. The issue is neither trivial nor unimportant since it can have a profound influence on the proof length complexity of the resulting tableau system. We will take up this discussion again in Chapter 6.1. There we will see that varying conditions (T3) and (T4) constitutes the general case of a technique which is commonly called *lemma generation* in the two-valued case.

A detailed derivation of the rule on the left in Figure 4.2 will further clarify the preceding definition.

Example 4.6. *Find the tableau rule for $\{\frac{1}{2}\}\,(\phi \vee \psi)$ in $\mathcal{L}_M^3$. The task is to find a minimal set of homomorphisms $\{h_1, h_2, \ldots\}$ whose corresponding rule extensions together cover all entries in the truth table of $\vee$ (see Table 2.1) that are equal to $\frac{1}{2}$. All h_i must satisfy (T1)–(T4). Obviously, it is sufficient to define h_i on ϕ, ψ (and, therefore, on all of their subformulas).*

$h_1(\phi) = \{\frac{1}{2}\}$, $h_1(\psi) = \{0, \frac{1}{2}\}$
defines the partial cover...

... adding the partial cover corresponding to $h_2(\phi) = \{0, \frac{1}{2}\}$ and $h_2(\psi) = \{\frac{1}{2}\}$ yields

ϕ/ψ	0	$\frac{1}{2}$	1
0	0	$\frac{1}{2}$	1
$\frac{1}{2}$	$\frac{1}{2}$	$\frac{1}{2}$	1
1	1	1	1

ϕ/ψ	0	$\frac{1}{2}$	1
0	0	$\frac{1}{2}$	1
$\frac{1}{2}$	$\frac{1}{2}$	$\frac{1}{2}$	1
1	1	1	1

It is easy to see that (T1)–(T4) are satisfied for h_1, h_2. And since the required cover cannot be achieved with a single homomorphism, (T0b) also holds.

Signed formulas for which no rule is defined are essentially self-contradictory formulas (see page 24) and have to be added to the contradiction set. Instead of pairs of complementary formulas we now have to look for finite sets of formulas with no common truth value that they can take on in a possible model.

Definition 4.7. (Contradiction set) *Given a logic $\mathcal{L}$ we define the* **contradiction set** *$Contr_S$ of many-valued tableau systems using sets of truth values as signs by*

$$Contr_S = \{\{S_1\,\phi, \ldots, S_r\,\phi\} \mid r \geq 1, \bigcap_{j=1}^r S_j = \emptyset, S_j\,\phi \in \mathbf{L}^*\}$$
$$\cup \{\{S\,\phi\} \mid \text{no rule defined for } S\,\phi \in \mathbf{L}^*\}$$

No other parts of the proof tree construction are affected by the new system. Note that in the above definition r is, in general, required to be greater than 2, for example when formulas such as $\{0, \frac{1}{2}\}\,\phi$, $\{0, 1\}\,\phi$,

$\{\frac{1}{2}, 1\}\ \phi$ are present on a branch. Let us abbreviate the fact that there is a closed proof tree over $N - D\ \phi$ with the string $\vdash_S \phi$.

At this point we may ask ourselves which signs should be chosen in order to produce an adequate tableau system. Certainly, not all of the 2^n possible ones (the size of $\bar{S}$) do make sense and these are also far too numerous to handle. We will give a very precise answer for a certain class of logics in the next chapter, but in general the following rule of thumb is helpful:

1. Include all signs necessary for a compact expression of your queries. If, for example, you are usually asking for logical consequence in the sense of Definition 2.21 then you are well advised to include D and $N - D$ in the set of signs. Call the resulting set of signs S_1.

2. Include all other signs that are necessary to write down short rules for the combinations of signs in S_1 and connectives in the logic.

3. Check that (4.2) below is satisfied. If not, add a sufficient selection of signs and goto 2.

Note that the signs $\{\}$ and N are not explicitly excluded, although a branch containing a formula such as $\{\}\ \phi$ can always be closed immediately, since no rule can be defined for such a formula* and a formula of the form $N\ \phi$ does not add any real information.

Example 4.8. *In Figures 4.3 and 4.4 we give the full tableau system for the propositional part of the logic from Example 4.2. Note that the rules for sign 0 with strong equivalence and affirmation of* $\{\frac{1}{2}\}$ *could have been simplified if the sign* $\{0, 1\}$ *had been available. It is instructive to compare the size of the rules in this system with the size of the rules that are generated by Surma and Carnielli's method described in Section 3.3.*

---

*By (T0a) and (T0b) only the empty set of extensions is possible, but then no homomorphism h exists and the rule is undefined.

Conjunction

$$\frac{\{0\}\,(\phi \wedge \psi)}{\{0\}\,\phi \mid \{0\}\,\psi} \qquad \frac{\{\frac{1}{2}\}\,(\phi \wedge \psi)}{\begin{array}{c|c} \{\frac{1}{2},1\}\,\phi & \{\frac{1}{2}\}\,\phi \\ \{\frac{1}{2}\}\,\psi & \{\frac{1}{2},1\}\,\psi \end{array}} \qquad \frac{\{1\}\,(\phi \wedge \psi)}{\begin{array}{c}\{1\}\,\phi \\ \{1\}\,\psi\end{array}}$$

$$\frac{\{0,\frac{1}{2}\}\,(\phi \wedge \psi)}{\{0,\frac{1}{2}\}\,\phi \mid \{0,\frac{1}{2}\}\,\psi} \qquad \frac{\{\frac{1}{2},1\}\,(\phi \wedge \psi)}{\begin{array}{c}\{\frac{1}{2},1\}\,\phi \\ \{\frac{1}{2},1\}\,\psi\end{array}}$$

Disjunction

$$\frac{\{0\}\,(\phi \vee \psi)}{\begin{array}{c}\{0\}\,\phi \\ \{0\}\,\psi\end{array}} \qquad \frac{\{\frac{1}{2}\}\,(\phi \vee \psi)}{\begin{array}{c|c} \{0,\frac{1}{2}\}\,\phi & \{\frac{1}{2}\}\,\phi \\ \{\frac{1}{2}\}\,\psi & \{0,\frac{1}{2}\}\,\psi \end{array}} \qquad \frac{\{1\}\,(\phi \vee \psi)}{\{1\}\,\phi \mid \{1\}\,\psi}$$

$$\frac{\{0,\frac{1}{2}\}\,(\phi \vee \psi)}{\begin{array}{c}\{0,\frac{1}{2}\}\,\phi \\ \{0,\frac{1}{2}\}\,\psi\end{array}} \qquad \frac{\{\frac{1}{2},1\}\,(\phi \vee \psi)}{\{\frac{1}{2},1\}\,\phi \mid \{\frac{1}{2},1\}\,\psi}$$

Strong negation

$$\frac{\{0\}\,\neg\phi}{\{1\}\,\phi} \qquad \frac{\{\frac{1}{2}\}\,\neg\phi}{\{\frac{1}{2}\}\,\phi} \qquad \frac{\{1\}\,\neg\phi}{\{0\}\,\phi} \qquad \frac{\{0,\frac{1}{2}\}\,\neg\phi}{\{\frac{1}{2},1\}\,\phi} \qquad \frac{\{\frac{1}{2},1\}\,\neg\phi}{\{0,\frac{1}{2}\}\,\phi}$$

Affirmation of $\{0\}$

$$\frac{\{0\}\,J_0(\phi)}{\{\frac{1}{2},1\}\,\phi} \quad \frac{\{\frac{1}{2}\}\,J_0(\phi)}{\text{No Rule}} \quad \frac{\{1\}\,J_0(\phi)}{\{0\}\,\phi} \quad \frac{\{0,\frac{1}{2}\}\,J_0(\phi)}{\{\frac{1}{2},1\}\,\phi} \quad \frac{\{\frac{1}{2},1\}\,J_0(\phi)}{\{0\}\,\phi}$$

Affirmation of $\{\frac{1}{2}\}$

$$\frac{\{0\}\,J_{\frac{1}{2}}(\phi)}{\{0\}\,\phi \mid \{1\}\,\phi} \quad \frac{\{\frac{1}{2}\}\,J_{\frac{1}{2}}(\phi)}{\text{No Rule}} \quad \frac{\{1\}\,J_{\frac{1}{2}}(\phi)}{\{\frac{1}{2}\}\,\phi} \quad \frac{\{0,\frac{1}{2}\}\,J_{\frac{1}{2}}(\phi)}{\{0\}\,\phi \mid \{1\}\,\phi} \quad \frac{\{\frac{1}{2},1\}\,J_{\frac{1}{2}}(\phi)}{\{\frac{1}{2}\}\,\phi}$$

Affirmation of $\{1\}$

$$\frac{\{0\}\,J_1(\phi)}{\{0,\frac{1}{2}\}\,\phi} \quad \frac{\{\frac{1}{2}\}\,J_1(\phi)}{\text{No Rule}} \quad \frac{\{1\}\,J_1(\phi)}{\{1\}\,\phi} \quad \frac{\{0,\frac{1}{2}\}\,J_1(\phi)}{\{0,\frac{1}{2}\}\,\phi} \quad \frac{\{\frac{1}{2},1\}\,J_1(\phi)}{\{1\}\,\phi}$$

FIG. 4.3. Tableau system for the logic from Example 4.2, Part I.

Weak negation

$$\frac{\{0\} \sim \phi}{\{1\}\, \phi} \qquad \frac{\{\frac12\} \sim \phi}{\text{No Rule}} \qquad \frac{\{1\} \sim \phi}{\{0,\frac12\}\, \phi} \qquad \frac{\{0,\frac12\} \sim \phi}{\{1\}\, \phi} \qquad \frac{\{\frac12,1\} \sim \phi}{\{0,\frac12\}\, \phi}$$

Implication

$$\frac{\{0\}\,(\phi \supset \psi)}{\begin{array}{c}\{1\}\,\phi \\ \{0\}\,\psi\end{array}} \qquad \frac{\{\frac12\}\,(\phi \supset \psi)}{\begin{array}{c}\{1\}\,\phi \\ \{\frac12\}\,\psi\end{array}} \qquad \frac{\{1\}\,(\phi \supset \psi)}{\{0,\frac12\}\,\phi \mid \{1\}\,\psi}$$

$$\frac{\{0,\frac12\}\,(\phi \supset \psi)}{\begin{array}{c}\{1\}\,\phi \\ \{0,\frac12\}\,\psi\end{array}} \qquad \frac{\{\frac12,1\}\,(\phi \supset \psi)}{\{0,\frac12\}\,\phi \mid \{\frac12,1\}\,\psi}$$

Weak equivalence

$$\frac{\{0\}\,(\phi \equiv \psi)}{\begin{array}{c|c}\{1\}\,\phi & \{0\}\,\phi \\ \{0\}\,\psi & \{1\}\,\psi\end{array}} \qquad \frac{\{\frac12\}\,(\phi \equiv \psi)}{\begin{array}{c|c}\{1\}\,\phi & \{\frac12\}\,\phi \\ \{\frac12\}\,\psi & \{1\}\,\psi\end{array}} \qquad \frac{\{1\}\,(\phi \equiv \psi)}{\begin{array}{c|c}\{0,\frac12\}\,\phi & \{1\}\,\phi \\ \{0,\frac12\}\,\psi & \{1\}\,\psi\end{array}}$$

$$\frac{\{0,\frac12\}\,(\phi \equiv \psi)}{\begin{array}{c|c}\{1\}\,\phi & \{0,\frac12\}\,\phi \\ \{0,\frac12\}\,\psi & \{1\}\,\psi\end{array}} \qquad \frac{\{\frac12,1\}\,(\phi \equiv \psi)}{\begin{array}{c|c}\{0,\frac12\}\,\phi & \{\frac12,1\}\,\phi \\ \{0,\frac12\}\,\psi & \{\frac12,1\}\,\psi\end{array}}$$

Strong equivalence

$$\frac{\{0\}\,(\phi \cong \psi)}{\begin{array}{c|c|c|c}\{\frac12,1\}\,\phi & \{1\}\,\phi & \{0\}\,\phi & \{0,\frac12\}\,\phi \\ \{0\}\,\psi & \{0,\frac12\}\,\psi & \{\frac12,1\}\,\psi & \{1\}\,\psi\end{array}} \qquad \frac{\{\frac12\}\,(\phi \cong \psi)}{\text{No Rule}}$$

$$\frac{\{1\}\,(\phi \cong \psi)}{\begin{array}{c|c|c}\{0\}\,\phi & \{\frac12\}\,\phi & \{1\}\,\phi \\ \{0\}\,\psi & \{\frac12\}\,\psi & \{1\}\,\psi\end{array}} \qquad \frac{\{0,\frac12\}\,(\phi \cong \psi)}{\begin{array}{c|c|c|c}\{\frac12,1\}\,\phi & \{1\}\,\phi & \{0\}\,\phi & \{0,\frac12\}\,\phi \\ \{0\}\,\psi & \{0,\frac12\}\,\psi & \{\frac12,1\}\,\psi & \{1\}\,\psi\end{array}}$$

$$\frac{\{\frac12,1\}\,(\phi \cong \psi)}{\begin{array}{c|c|c}\{0\}\,\phi & \{\frac12\}\,\phi & \{1\}\,\phi \\ \{0\}\,\psi & \{\frac12\}\,\psi & \{1\}\,\psi\end{array}}$$

Fɪɢ. 4.4. Tableau system for the logic from Example 4.2, Part II.

We close this section with an example of a proof tree constructed using the system from Figures 4.3 and 4.4.

Example 4.9. *We show that $\vdash_S \neg p \supset (\sim p \wedge \neg p)$ holds in the tableau system corresponding to the logic from Example 4.2. The same fact was proven in Example 3.17. Note that the single tree required now is exactly the size of the smaller one of the two trees before.*

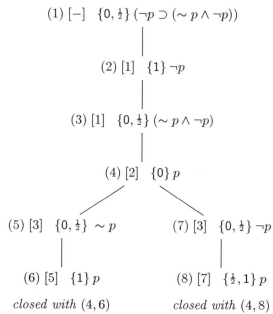

$$(1)\ [-]\quad \{0, \tfrac{1}{2}\}\,(\neg p \supset (\sim p \wedge \neg p))$$

$$(2)\ [1]\quad \{1\}\,\neg p$$

$$(3)\ [1]\quad \{0, \tfrac{1}{2}\}\,(\sim p \wedge \neg p)$$

$$(4)\ [2]\quad \{0\}\,p$$

$$(5)\ [3]\quad \{0, \tfrac{1}{2}\}\,\sim p \qquad\qquad (7)\ [3]\quad \{0, \tfrac{1}{2}\}\,\neg p$$

$$(6)\ [5]\quad \{1\}\,p \qquad\qquad\quad (8)\ [7]\quad \{\tfrac{1}{2}, 1\}\,p$$

$$\text{\textit{closed with }} (4, 6) \qquad\qquad \text{\textit{closed with }} (4, 8)$$

From now on we refer to the tableau systems introduced in this section as the **sets-as-signs approach** to discern it, for example from classical tableaux or Carnielli's rules.

4.2   Soundness

Before we proceed to prove soundness we state precisely which conditions should be imposed on the set of signs in order to guarantee it. Consider the tableau system for the logic $\mathcal{L}^3_{M+}$ from the previous section, but equipped with the set of signs $\{\{0\}, \{1\}\}$. As a consequence, the rules corresponding to formulas such as $\{\tfrac{1}{2}\}\,\neg\phi$ are not defined. Thus a branch containing such a formula could be closed, which is obviously unsound. To guarantee soundness we must therefore impose a technical condition on the set of signs so that 'enough' signs are present. Otherwise, not all rules that should be defined are actually defined, as we have just seen.

Definition 4.10. (Complete set of signs) *Let f be the interpretation of an m-place connective F and $\mathbf{S}$ the set of signs of an n-valued logic $\mathcal{L}$. $\mathbf{S}$ is called **complete** wrt $\mathcal{L}$ iff for all F in $\mathcal{L}$ and $\mathsf{S} \in \mathbf{S}$*

For all $(j_1, \ldots, j_m) \in f^{-1}(\mathsf{S})$,
  there exist $\mathsf{S}_1, \ldots, \mathsf{S}_m \in \mathbf{S} \cup \{N\}$
  such that $j_l \in \mathsf{S}_l, 1 \leq l \leq m$ *and* $f'(\mathsf{S}_1, \ldots, \mathsf{S}_m) \subseteq \mathbf{S}.$ (4.2)

From now on it will be assumed that all logics come with complete sets of signs. It is easy to prove that (4.1) represents an easy-to-check criterion for the completeness of the sets of signs in the sense defined above. The relationship between complete sets of signs and definiteness of rules is expressed in the following lemma.

Lemma 4.11. *Let* $\mathcal{L}$, f, *and* F *be as before and* $\mathbf{S}$ *complete. Then for any signed formula* $\mathsf{S}\, F(\phi_1, \ldots, \phi_m) \in \mathbf{L}^*$:

If $f^{-1}(\mathsf{S}) \neq \emptyset$ *then a tableau rule for* $\mathsf{S}\, F(\phi_1, \ldots, \phi_m)$ *is defined.*

Proof. We construct a set of homomorphisms $g_1, \ldots, g_r$ that satisfy (T0a), (T1), and (T2). Then, by Proposition 4.5, we know that there are also homomorphisms $h_1, \ldots, h_q$ that satisfy (T0)–(T4) and hence the rule is defined.

$f^{-1}(\mathsf{S}) \neq \emptyset$, say $f^{-1}(\mathsf{S}) = \{\vec{\jmath}^1, \ldots, \vec{\jmath}^r\}$, where the $\vec{\jmath}^k$ are m-tuples of truth values. Let $\vec{\mathsf{S}}^1, \ldots, \vec{\mathsf{S}}^r$ be tupels of signs $\vec{\mathsf{S}}^k = (\mathsf{S}_1^k, \ldots, \mathsf{S}_m^k)$ that satisfy (4.2) for each $\vec{\jmath}^k$, $1 \leq k \leq r$. We define g_k for $1 \leq k \leq r$ by

$$g_k(\phi_i) = \mathsf{S}_i^k \text{ if } \mathsf{S}_i^k \neq N$$

which gives us (T1). By (4.2) each $\vec{\jmath}^k \in f^{-1}(\mathsf{S})$ is covered by $\vec{\mathsf{S}}^k$, hence (T0a) also holds. (T2) also follows directly from (4.2), since for each $\vec{\mathsf{S}}^k$ it assures that $f'(\vec{\mathsf{S}}^k) \subseteq \mathbf{S}$.

Definition 4.12. (Signed formula and tableau satisfiability)   *Let* Φ *be a set of signed formulas.* Φ *is* **satisfiable** *in a logic* $\mathcal{L}$ *iff there is a* $\mathcal{L}$-*valuation* v *such that for all* $\mathsf{S}\, \phi \in \Phi$ *we have* $v(\phi) \in S$. *We say that* v *is a* **model** *for* Φ. *A tableau branch* $\mathbf{B}$ *is* **satisfiable** *iff its set of node labels is. A tableau* $\mathbf{T}$ *is* **satisfiable** *iff it contains at least one satisfiable branch.*

Next we have the usual lemma on preservation of satisfiability.

Lemma 4.13. *Let* $\mathbf{T}$ *be a satisfiable tableau and suppose* $\mathbf{T}'$ *was created by a rule application to an arbitrary formula in* $\mathbf{T}$. *Then* $\mathbf{T}'$ *is also satisfiable.*

Proof. $\mathbf{T}$ contains at least one satisfiable branch $\mathbf{B}$. If the formula in the rule application was not in $\mathbf{B}$ then $\mathbf{B}$ is unchanged and hence is still satisfiable.

On the other hand, let $\mathsf{S}\, F(\phi_1, \ldots, \phi_m)) \in \mathbf{B}$ be the formula which supplied the premise for the rule application and let v be a valuation that models $\mathbf{B}$. For such a valuation, by definition, $v(F(\phi_1, \ldots, \phi_m)) \in S$. Since v is a homomorphism, we have

$$v(F(\phi_1, \ldots, \phi_m)) = f(v(\phi_1), \ldots, v(\phi_m)) \in S.$$

Let $\mathsf{S}_1 \, \phi_{i_1} \circ \cdots \circ \mathsf{S}_t \, \phi_{i_t}$ be an extension obtained by using (T0a) on $(v(\phi_1), \ldots, v(\phi_m)) \in f^{-1}(\mathsf{S})$. Take any i_k $(1 \leq k \leq t)$: By (T0a) we have $v(\phi_{i_k}) \in S_k$. Together with the assumption that v is a model for **B** we have the satisfiability of $\mathbf{B} \cup \{\mathsf{S}_1 \, \phi_{i_1}, \ldots, \mathsf{S}_t \, \phi_{i_t}\}$, which concludes the proof.

Theorem 4.14. (Soundness) *Let ϕ be a $\mathcal{L}$-formula. If there is a closed tableau with root $\mathsf{N} - \mathsf{D} \, \phi$ then ϕ is a $\mathcal{L}$-tautology. In short, for any logic $\mathcal{L}$ and any $\phi \in \mathbf{L}$:*

$$If \vdash_\mathsf{S} \phi \ then \vDash_\mathcal{L} \phi$$

Proof. Let **T** be a closed tableau for $\mathsf{N} - \mathsf{D} \, \phi$. **T** cannot be satisfiable. Assume that **B** is an arbitrary branch in **T**. Since **T** is closed, there is a $C \subseteq \mathbf{B}$ with $C \in \mathbf{Contr}_\mathsf{S}$. If the first part of $\mathbf{Contr}_\mathsf{S}$ applies then there is no valuation that satisfies all formulas in C simultaneously. If the second part applies then $C = \{\mathsf{S} \, \phi\}$ and no rule is defined for $\mathsf{S} \, \phi$, hence by Lemma 4.11, $f^{-1}(\mathsf{S}) = \emptyset$ and no homomorphism $w : \mathbf{L} \to N$ such that $w(\phi) \in S$ and which is also defined on all subformulas of ϕ can exist. In both cases there can be no valuation that satisfies C and thus no valuation satisfies **B**. This holds for arbitrary branches, so **T** is not satisfiable.

The next step is to show by a straightforward induction, using Lemma 4.13, that any tableau with satisfiable root must itself be satisfiable.

We conclude that if **T** is not satisfiable then $\mathsf{N} - \mathsf{D} \, \phi$ is not satisfiable, which means by definition for all valuations v that $v(\phi) \notin N - D$ iff $v(\phi) \in D$ iff ϕ is a tautology.

4.3   Completeness

The completeness proof for our system is quite straightforward and closely follows the lines of standard tableau completeness proofs as, for instance, in Fitting (1990b), but in order to be able to deal with generalized signs we will have to make appropriate modifications to the definitions of a Hintikka Set and of the Analytic Consistency Property. We then proceed as usual, first proving Hintikka's lemma, and second a model existence theorem, which in turn yields completeness. For the sake of modularity and flexibility we prefer the formulation with analytic consistency properties over a more direct one. It is then easy to extend the proofs to first-order formulas or infinite sets of formulas. Also, other standard meta-results such as strong completeness (for suitable logics) and compactness may be easily obtained, although we do not include them here.

Definition 4.15. (Many-valued propositional Hintikka set)   *A set H of $\mathsf{S}_\mathcal{L}$-signed formulas is called a* **Hintikka set** *if it is contradiction free and downward saturated, or, more precisely, if the following two conditions hold:*

(H1) *For all signed atomic formulas* $S_i\, p \in L_0^*$: *If* $S_1\, p, \ldots, S_r\, p \in H$ *then* $\bigcap_{j=1}^{r} S_j \neq \emptyset$.

(H2) *If* $S\, F(\phi_1, \ldots, \phi_m) \in H$ *then* $\pi_{S,F}$ *is defined and at least one of the extensions* *determined hereby* $\{S_1\, \phi_{i_1}, \ldots, S_t\, \phi_{i_t}\}$ *is a subset of* H.

A Hintikka set $\bar{H}$ *is called a* **saturated Hintikka set** *or* **model set** *if in addition to the above stated conditions it is atomically complete and upward saturated, that is, if the following hold:*

(H3) *For all propositional variables* $p \in L_0$ *there exists an* $S \in S_{\mathcal{L}}$ *such that* $S\, p \in \bar{H}$.

(H4) *If* $S \in S_{\mathcal{L}}$ *then* $S\, F(\phi_1, \ldots, \phi_m) \in \bar{H}$, *whenever at least one of the extensions* $\{S_1\, \phi_{i_1}, \ldots, S_t\, \phi_{i_t}\}$ *determined by* $\pi_{S,F}$ *is a subset of* $\bar{H}$.

Note that by (H1) and (H2) it is impossible that $\{\}\, \phi$ for any $\phi \in \mathbf{L}$ ever occurs in a (saturated) Hintikka set. It may seem strange at first sight that (H3) would even be satisfied if $N\, p \in \bar{H}$ for some p, because $N\, p$ does not give any information about the truth value of p. The definition becomes more natural if we remember that, also in the proof of Hintikka's lemma in the classical version, the truth value of the atoms that are being added is completely arbitrary; it is only for definiteness that they are assigned a truth value. This is exactly what the formula $N\, p \in \bar{H}$ says, that p has a definite truth value, but we do not care which one.

Lemma 4.16. (Hintikka's lemma) *Every Hintikka set* H *can be extended to a saturated Hintikka set* $\bar{H}$.

Proof. Let H be a Hintikka set, $L_0 = \{p_i \mid i \in \mathbb{N}\}$ an enumeration of the propositional variables, and $S_0 \in S_{\mathcal{L}}$ arbitrary. We extend H to a saturated Hintikka set $\bar{H}$ in the following way:

$$
\begin{aligned}
H_0 \;&=\; H \cup \{S_0\, p_i \mid p_i \in L_0 \text{ and } S\, p_i \notin H \text{ for all } S \in S_{\mathcal{L}}\} \\
H_{i+1} \;&=\; H_i \cup \\
&\quad \{S\, F(\phi_1, \ldots, \phi_m) \mid S \in S_{\mathcal{L}}, \pi_{S,F} \text{ defined and at least one of} \\
&\qquad \text{the extensions } \{S_1\, \phi_{i_1}, \ldots, S_t\, \phi_{i_t}\} \text{ deter-} \\
&\qquad \text{mined by } \pi_{S,F} \text{ is a subset of } H_i\} \\
\bar{H} \;&=\; \bigcup_{i \in \mathbb{N}} H_i
\end{aligned}
$$

First we extend H so that it assigns a definite truth value set[†] to each variable not already occurring in H; then we inductively take all formulas with a greater depth that follow into account.

---

*Here and in the following we treat extensions as sets.

[†]We can take any $S_0 \in S_{\mathcal{L}}$, for example N would have done as well provided it was in $S_{\mathcal{L}}$. See also the remark preceding Hintikka's lemma.

(H1) holds for $\bar{H}$, for, let $p \in L_0$, then either there exists a $\mathsf{S} \in \mathsf{S}_\mathcal{L}$ such that $\mathsf{S}\, p \in H$, nothing is changed by the construction, and (H1) still holds, or $\mathsf{S}\, p \notin H$ for all $\mathsf{S} \in \mathsf{S}_\mathcal{L}$, then $\mathsf{S}_0\, p$ is added and since this is the only atomic occurrence of p in $\bar{H}$, (H1) holds trivially.

(H3) and (H4) hold by construction of $\bar{H}$.

To see that (H2) holds, let $\mathsf{S}\, F(\phi_1, \ldots, \phi_m) \in \bar{H}$ and $\mathsf{S} \in \mathsf{S}_\mathcal{L}$. Then either, already, $\mathsf{S}\, F(\phi_1, \ldots, \phi_m) \in H$ and (H2) is inherited from H or $\mathsf{S} F(\phi_1, \ldots, \phi_m)$ was generated during the construction in some H_i for $i > 0$. Then, by definition, $\pi_{\mathsf{S},F}$ is defined and at least one of the extensions $\{\mathsf{S}_1\, \phi_{i_1}, \ldots, \mathsf{S}_t\, \phi_{i_t}\}$ determined by $\pi_{\mathsf{S},F}$ is in H_{i-1} and (H2) is inherited from H_{i-1}.

Definition 4.17. (Subset closed, finite character) *Let Γ be a family of sets. Γ is called **subset closed** if $K \in \Gamma$ implies that $K' \in \Gamma$ for any $K' \subseteq K$. Γ has **finite character** if a set K belongs to Γ iff all finite subsets of K belong to Γ.*

Definition 4.18. (Analytic consistency property) *A family Γ ranging over sets of $\mathsf{S}_\mathcal{L}$-signed formulas is called the **Analytic Consistency Property (ACP)** iff for all $K \in \Gamma$ the following conditions hold:*

(ACP1) *For all propositional variables $p \in L_0$: If $\mathsf{S}_1\, p, \ldots, \mathsf{S}_r\, p \in K$ then $\bigcap_{j=1}^{r} \mathsf{S}_j \neq \emptyset$.*
(ACP2) *If $\mathsf{S}\, F(\phi_1, \ldots, \phi_m) \in K$ then $\pi_{\mathsf{S},F}$ is defined and for at least one of the hereby determined extensions $C = \{\mathsf{S}_1\, \phi_{i_1}, \ldots, \mathsf{S}_t\, \phi_{i_t}\}$ we have $K \cup C \in \Gamma$.*

Note that (ACP1) is identical to (H1) and (ACP2) is closely related to (H2). Thus an ACP can be regarded as a sequence of approximations of a Hintikka set, and as it turns out, it always actually contains a Hintikka set.

If $K \in \Gamma$ then K is called Γ-**consistent**. Every ACP Γ of finite character by a standard argument is also subset closed.

Theorem 4.19. (Model existence) *Let Γ be an ACP of finite character and K a Γ-consistent set, that is, $K \in \Gamma$. Then there exists a valuation v, such that $v(\phi) \in \mathsf{S}$ holds whenever $\mathsf{S}\, \phi \in K$; in other words, v is a model for K.*

Proof. In a first step we will extend K with a Lindenbaum construction restricted to ACPs in order to find a $\mathbf{L}^*$-maximal element M in Γ (this step corresponds to Tukey's lemma in the denumerable case); in a second step we show that M is a Hintikka set, so that we can use it to define an appropriate valuation.

Let $\{\zeta_1, \zeta_2, \ldots\}$ be an enumeration of all signed formulas in $\mathbf{L}^*$ and define C_n for $n \geq 0$ as follows:

$$C_0 = K$$
$$C_{n+1} = \begin{cases} C_n \cup \{\zeta_n\} & \text{if } C_n \cup \{\zeta_n\} \in \Gamma \\ C_n & \text{otherwise} \end{cases}$$

Clearly, all C_n are members of Γ and, ordered by set inclusion, constitute a chain in Γ. We define

$$M = \bigcup_{n \geq 0} C_i$$

and thus have:

1. M is $\mathbf{L}^*$-maximal in Γ, since

   (**Membership**) Let $M' \subseteq M$ be arbitrary, but finite. Hence we have some C_n with $M' \subseteq C_n$ (take as n the highest index of all elements in M') and from $C_n \in \Gamma$, we also have that $M' \in \Gamma$ because Γ is subset closed. Thus $M' \in \Gamma$ for all finite $M' \subseteq M$ and hence $M \in \Gamma$ because of the finite character of Γ.

   (**Maximality**) Assume that there is $M' \subseteq \mathbf{L}^*$ with $M \subsetneq M' \in \Gamma$. Then there must exist some $\zeta_n \in M'$ with $\zeta_n \notin M$. By definition, we have $C_n \subseteq M \subsetneq M' = M' \cup \{\zeta_n\}$, hence $C_n \cup \{\zeta_n\} \subseteq M'$. Since Γ is subset closed we know that $C_n \cup \{\zeta_n\} \in \Gamma$. But, by definition, $C_n \cup \{\zeta_n\} = C_{n+1} \subseteq M$ thus, already, $\zeta_n \in M$, which is a contradiction.

2. Membership and (ACP1) imply (H1), maximality and (ACP2) imply (H2) for M, so M is indeed a Hintikka set. According to Hintikka's lemma we can extend M to a saturated Hintikka set $\bar{M}$.

It remains to show that $\bar{M}$ determines a model for K. For this purpose we fix an arbitrary function v that obeys

$$v(p) \in S \text{ iff } \mathsf{S}\, p \in \bar{M}$$

for all $p \in L_0$. We show that v determines a valuation that models $\bar{M}$. $\bar{M}$ is a saturated Hintikka set. (H1) guarantees that v is well-defined and (H3) that it is total on L_0.

We extend v to a homomorphism from $\mathbf{L}$ to $\mathbf{A}$ and show by induction on the depth of formulas ϕ that $\phi \in L$ and $\mathsf{S}\,\phi \in \bar{M}$ imply $v(\phi) \in S$.[*]

The case when ϕ is atomic is settled by definition.

Suppose that $\phi = F(\phi_1, \ldots, \phi_m)$. According to (H2) there is at least one extension determined by $\pi_{S,F}$ with $\{\mathsf{S}_1\,\phi_{i_1}, \ldots, \mathsf{S}_t\,\phi_{i_t}\} \subseteq \bar{M}$. The

---

[*]In the proof we do not make use of (H4). In fact, using (H4) we could also show the other direction, namely that for any $\phi \in \mathbf{L}$ there exists a $\mathsf{S} \in \mathsf{S}_{\mathcal{L}}$ such that $v(\phi) \in S$ implies $\mathsf{S}\,\phi \in \bar{M}$.

induction hypothesis yields $v(\phi_{i_k}) \in S_k$ for $1 \leq k \leq t$. Hence we can conclude

$$
\begin{aligned}
v(\phi) &= v(F(\phi_1, \ldots, \phi_m)) \\
&= f(v(\phi_1), \ldots, v(\phi_m)) && (v \text{ homomorphism}) \\
&\in f'(\{v(\phi_1)\}, \ldots, \{v(\phi_m)\}) && (\text{definition of } f') \\
&= f'(\{v(\phi_1)\}, \ldots, \{v(\phi_{i_1})\}, \ldots, \{v(\phi_{i_t})\}, \ldots, \{v(\phi_m)\}) \\
&\subseteq f'(\mathsf{N}, \ldots, \mathsf{S}_1, \ldots, \mathsf{S}_t, \ldots, \mathsf{N}) && (\text{induction hypothesis}) \\
&\subseteq \mathsf{S} && (\text{Definition 4.4, T2})
\end{aligned}
$$

So we have indeed constructed a model for $\bar{M}$ and the theorem follows from the fact that $K \subseteq \bar{M}$.

Lemma 4.20.

1. *Any ACP can be extended to one that is subset closed.*
2. *Any ACP that is subset closed can be extended to one that is of finite character.*

Proof.

1. Let Γ be an ACP. Define $\Gamma^+ = \{K \mid K \subseteq K' \in \Gamma\}$. Clearly, $\Gamma \subseteq \Gamma^+$ and Γ^+ is subset closed. It remains to show that Γ^+ is an ACP. Let $K \in \Gamma^+$ arbitrary. (ACP1) holds for K, since $K \subseteq K'$ for some $K' \in \Gamma$ and (ACP1) is inherited from Γ. For (ACP2) assume that $\mathsf{S}\, F(\phi_1, \ldots, \phi_m) \in K$. Hence, $\mathsf{S}\, F(\phi_1, \ldots, \phi_m) \in K'$ for some $K' \in \Gamma$ and since Γ is an ACP we have that a rule is defined and $K' \cup C \in \Gamma$ for some extension C. But then, by definition of Γ^+ and $K \cup C \subseteq K' \cup C$, $K \cup C \in \Gamma^+$ also.

2. Let Γ be an ACP that is subset closed. Define

$$\Gamma^+ = \{K \mid K' \in \Gamma \text{ for all finite subsets } K' \text{ of } K\}.$$

Since Γ is subset closed for each $K' \in \Gamma$ all (finite) subsets are in Γ, hence $\Gamma \subseteq \Gamma^+$ and, obviously, Γ^+ is of finite character. If (ACP1) would not hold for some $K \in \Gamma^+$ there would be a finite set $\{\mathsf{S}_1\, p, \ldots, \mathsf{S}_r\, p\} \subseteq K$ for that $\bigcap_{j=1}^r \mathsf{S}_j = \emptyset$. This set is, by definition, in Γ, thus causing a contradiction. For (ACP2) assume that $\mathsf{S}F(\phi_1, \ldots, \phi_m) \in K \in \Gamma^+$. First observe that a rule must be defined, since $\{\mathsf{S}\, F(\phi_1, \ldots, \phi_m)\} \in \Gamma$. If K is finite it must also be in Γ by construction and we are done. If K is infinite, consider the chain

$$\{\mathsf{S}\, F(\phi_1, \ldots, \phi_m)\} \subseteq \cdots \subseteq \{\mathsf{S}\, F(\phi_1, \ldots, \phi_m), \kappa_1, \ldots, \kappa_i\} \subseteq \cdots \subseteq K$$

for an arbitrary enumeration $\{\kappa_1, \kappa_2, \ldots\}$ of K. Each of the members of the chain is in Γ and defines some extension C_i of $\pi_{\mathsf{S},F}$ such that $\{\mathsf{S}\, F(\phi_1, \ldots, \phi_m), \kappa_1, \ldots, \kappa_i\} \cup C_i \in \Gamma$. Since $\pi_{\mathsf{S},F}$ has a finite number of extensions, at least one extension occurs infinitely often. Let C be such an extension. It suffices to show that $K' \in \Gamma$ for each finite subset K' of $K \cup C$. Since C occurs infinitely often for each K' there exists a finite set $K'' \in \Gamma$ with $K' \subseteq K'' \subseteq K \cup C$ and $K'' \cup C \in \Gamma$. From the fact that Γ is subset closed we have the desired result, namely $K' \in \Gamma$.

Theorem 4.21. (Completeness) *Let ϕ be a $\mathcal{L}$-formula. If ϕ is a $\mathcal{L}$-tautology then there is a closed tableau with root $\mathsf{N} - \mathsf{D}\,\phi$. In short, for any logic $\mathcal{L}$ and any $\phi \in \mathbf{L}$:*

$$\text{If } \vDash_{\mathcal{L}} \phi \text{ then } \vdash_{\mathsf{S}} \phi$$

Proof. Since ϕ is a tautology, for all valuations $v(\phi) \in D$ must hold. Suppose no closed tableau with root $\mathsf{N} - \mathsf{D}\,\phi$ exists. It follows that there exists at least one complete tableau $\mathbf{T}$ with root $\mathsf{N} - \mathsf{D}\,\phi$ that contains a complete open branch $\mathbf{B_T}$.

Define $\mathcal{B}$ as the collection of all finite $\mathcal{L}$-tableau branches that cannot be closed. For all $\mathbf{B} \in \mathcal{B}$ we have:

- For all propositional variables $p \in L_0$, if $\mathsf{S}_1\, p, \ldots, \mathsf{S}_r\, p \in \mathbf{B}$ then $\bigcap_{j=1}^{r} \mathsf{S}_j \neq \emptyset$, otherwise $\mathbf{B}$ would be closed.
- If $\mathsf{S}\, F(\phi_1, \ldots, \phi_m) \in \mathbf{B}$ then $\pi_{\mathsf{S},F}$ is defined and for at least one of the extensions determined hereby $C = \{\mathsf{S}_1\, \phi_{i_1}, \ldots, \mathsf{S}_t\, \phi_{i_t}\}$ we have $\mathbf{B} \cup C \in \mathcal{B}$. For, assume that $\pi_{\mathsf{S},F}$ was not defined, then $\mathbf{B}$ would be closed. On the other hand, if for no C we had $\mathbf{B} \cup C \in \mathcal{B}$ then for *all* C $\mathbf{B} \cup C$ could be closed and so could $\mathbf{B}$, a contradiction.

Putting these facts together, we have that $\mathcal{B}$ is an ACP; moreover, $\mathsf{N} - \mathsf{D}\,\phi$ is $\mathcal{B}$-consistent, because $\mathsf{N} - \mathsf{D}\,\phi \in \mathbf{B_T} \in \mathcal{B}$.

By Lemma 4.20 we can extend $\mathcal{B}$ to an ACP of finite character. From the Model Existence Theorem 4.19, we know that there exists a valuation v with $v(\phi) \in N - D$ and this is the contradiction we have been looking for.

We promised to point out what has to be changed in order to handle 0-ary connectives. These are best treated as atomic formulas, although syntactically, they are formulas of depth 1. So tableau rules are never applied to 0-ary connectives. Their semantics is covered by extending the contradiction set. Consequently, the base cases for the Hintikka set and ACP must also be extended and so must the proofs of Theorem 4.14, Lemma 4.16, Theorem 4.19, Lemma 4.20, and Theorem 4.21. The changes, however, are all straightforward to implement.

4.4   Size of proof trees

In this section we take a look at the size of proof trees in order to gain some measure of what has been achieved so far. Since the *depth* of proof trees in the propositional case depends only on the depth of the formulas to be proved we concentrate on their *fatness* rather than on their depth. In the following we again take up the considerations on average branching factors of rules made at the end of Section 3.3.

First we can state that our system is exponentially better than Surma and Carnielli's system. This can be seen already through trivial examples. Consider the three-valued tautology

$$((\cdots (p_1 \vee p_2) \vee \cdots \vee p_m) \vee \sim p_1)$$

It is an easy exercise to convince oneself that its shortest proof tree has exactly one branch and $2m + 2$ nodes. The two proof trees required when Surma and Carnielli's method is used, however, always have $\mathcal{O}(2^m)$ size. When $n > 3$ it even becomes $\mathcal{O}(2^{mn})$, while the proof tree obtained with our method still has $2m + 2$ nodes.*

In the following let us concentrate on two-place connectives.

The worst-case branching factor per rule for Surma and Carnielli's method is $n^2 - n$, while the average branching factor is n. We compare this with our system.

Proposition 4.22. *For any logic $\mathcal{L}$, if $\mathbf{S}_{\mathcal{L}}$ contains a sufficient number of signs then no rule constructed according to Definition 4.4 has more than n extensions with at most two formulas in each.*

Proof. Let F be any two-place connective and S any sign. Consider the n-branching rule for S $F(\phi_1, \phi_2)$ whose i-th extension ($i \in |N|$) is defined as

$$\left\{ \frac{i}{n-1} \right\} \phi_1 \circ \mathsf{S}_i \ \phi_2, \text{ where } \mathsf{S}_i = \left\{ \frac{j}{n-1} \middle| \ f(i,j) \in S \right\}$$

If we define

$$h_i(\phi_k) = \begin{cases} \{ \frac{i}{n-1} \} & k = 1 \\ \mathsf{S}_i & k = 2 \end{cases}$$

then it is easy to see that (T0a, T1, T2) are satisfied when we use the homomorphisms $h_0, \ldots, h_{n-1}$. The rest follows by Proposition 4.5 and by observing that each extension contains at most two formulas for a binary connective by definition.

This result ensures that our rules cannot grow too 'bulky'; however, n-valued strong equivalence (cf. Definition 2.29) shows that the bounds can actually be reached. The rule for $\{1\} \ (\phi \cong \psi)$ is

---

*It is instructive to compute the example for $n = m = 3$.

$$\frac{\{1\}\,(\phi \cong \psi)}{\begin{array}{c|c|c|c}
\{0\}\,\phi & \{\frac{1}{n-1}\}\,\phi & & \{1\}\,\phi \\
\{0\}\,\psi & \{\frac{1}{n-1}\}\,\phi & \cdots & \{1\}\,\psi
\end{array}}$$

The rules for most other signs have a smaller branching factor. There are logics, however, in which n extensions per rule are reached for *each* combination of sign and connective and hence have a branching factor of n *on average*.

Such an operator $\oplus$ could be defined as

$$i \oplus j = \frac{((i+j)\cdot(n-1)) \bmod n}{n-1}$$

for which we have that $\oplus^{-1}(k) = P_1(k) \cup P_2(k)$, where

$$P_1(k) = \{(i,j)|i+j=k\}$$
$$P_2(k) = \{(i,j)|i+j=k+1\}$$

It is easy to see that for all k

$$(i_1,j_1) \in P_1(k), (i_2,j_2) \in P_2(k) \text{ implies } i_1 < i_2, j_1 < j_2$$

which yields the branching property if one considers the diagonal shape of the $P_i(k)$.

The conclusion to be drawn from all this is that while our approach achieves a substantial improvement for many common many-valued logics and, in general, improves the worst-case branching factor, there are still some logics for which there is little or no improvement. While in logics such as $\mathcal{L}_M^n$ proofs are not much longer than their classical counterparts*, proofs in n-valued Łukasiewicz logic, for example, already become intractable, for small n.

Clearly, the method we have developed so far is not entirely satisfactory in all cases, particularly as operators such as strong equivalence, Łukasiewicz implication, and $\oplus$ have simple definitions, the former being very common. It turns out, however, that there is a nice solution to this problem, which we will discuss in Chapter 6.

On the other hand, the worst-case is never reached, for example, in the system for $\mathcal{L}_M^n$ described in Example 4.2. This behaviour is typical for many logics found in the literature.

*Since all many-valued tautologies are also classical tautologies one can make comparisons on this basis.

4.5 Function minimization

Assume that we are working in the logic from Example 4.2, but with the additional sign $\{0,1\}$ present. The definition of tableau rules admits the following two different rules for the signed formula $\{0\}\,(\phi \cong \psi)$:

$$
\begin{array}{c}
\{0\}\,(\phi \cong \psi) \\ \hline
\begin{array}{c|c|c}
\{\frac{1}{2},1\}\,\phi & \{0,1\}\,\phi & \{0,\frac{1}{2}\}\,\phi \\
\{0\}\,\psi & \{\frac{1}{2}\}\,\psi & \{1\}\,\psi
\end{array}
\end{array}
\qquad
\begin{array}{c}
\{0\}\,(\phi \cong \psi) \\ \hline
\begin{array}{c|c|c}
\{1\}\,\phi & \{\frac{1}{2}\}\,\phi & \{0\}\,\phi \\
\{0,\frac{1}{2}\}\,\psi & \{0,1\}\,\psi & \{\frac{1}{2},1\}\,\psi
\end{array}
\end{array}
$$

This kind of indeterminism stems from certain symmetries in the definition of $\cong$ and can hardly be avoided in efficient representations such as the one we use.

This becomes quite clear as soon as one notes that there is a close connection between our definition of tableau rules and work done in the area of (many-valued) function minimization (FM), where similar effects can be observed.

In the FM literature, the representation of a logical operator with a disjunction of conjunctions (which is exactly what our tableau rules are) is usually called a **sum-of-products** (SOP) expression. There, the aim is to find minimal SOP representations of many-valued logical functions, and this, in a sense, is exactly what we do when we compute tableau rules. In fact, it is sufficient to minimize *two-valued* operators for our purposes.

In the FM area, as a tool for finding minimal representations, so-called *Karnaugh maps* (Karnaugh, 1953) have been introduced. With Karnaugh maps it is possible to visualize k-ary operators for $k > 2$. Consider, for example, the two-valued four-place function*

$$f(x_1,x_2,x_3,x_4) = (\neg x_1 \wedge \neg x_3) \vee (\neg x_2 \wedge x_3 \wedge x_4) \vee (x_1 \wedge x_2 \wedge x_3) \vee (x_2 \wedge x_3 \wedge \neg x_4)$$

which can be conveniently represented as in the diagram on the left in Figure 4.1. The value of f for a vector $\vec{x} = (x_1,x_2,x_3,x_4)$ can be found in the entry corresponding to the row labelled with the values of $x_3 x_4$ and the column labelled with the values of $x_1 x_2$. For example, $f(0,0,1,0) = 0$. We now need some FM terminology.

Definition 4.23. (Minterm) *An element of the domain of a k-place many-valued connective f, in other words, an element of $f^{-1}(N) \subseteq N^k$, is called a* **minterm** *of f.*

Definition 4.24. (ON-set) *The* **ON-set** *of a two-valued connective f is the set of all minterms $\vec{x}$ such that $f(\vec{x}) = 1$.*

Definition 4.25. (Literal operator) *The* **literal operator** $[\,\cdot\,]^S$ *is a unary propositional connective which is defined for $S \subseteq N$ as follows:*

---

*The example is taken from Dueck (1988, p. 12f); this reference can also serve as a pointer to further FM literature. Note that there are some errors in Dueck (1988, p. 12, (i)–(iv)) which have been corrected in our example.

$$i^S = \begin{cases} 1 & i \in S \\ 0 & i \notin S \end{cases}$$

If S is a singleton write ϕ^i instead of $\phi^{\{i\}}$. For two-valued functions write ϕ instead of ϕ^1 and $\bar{\phi}$ instead of ϕ^0.

A *literal* in the present setting is a literal operator applied to a propositional variable.

Definition 4.26. (Product term, sum-of-products)   *A* **product term** *is a conjunction of literals, usually written as $x_1^{S_1} x_2^{S_2} \cdots x_r^{S_r}$. A* **sum-of-products** *is a disjunction of product terms, usually written as $p_1 + \cdots + p_l$, where the p_i are product terms.*

Definition 4.27. (Implicant, prime implicant) *An* **implicant** *for a two-valued function f is a product term whose ON-set is a subset of the ON-set of f. A* **prime implicant** *of f is an implicant that is not properly contained in any other implicant of f. An* **essential prime implicant** *of f is a prime implicant whose ON-set contains a minterm that does not occur in the ON-set of any other prime implicant.*

We illustrate the definitions with the example from above.

Some minterms of f, all of which are in the ON-set of f, are $(0,0,0,0)$, $(0,1,0,1)$, $(1,1,1,0)$. A sum-of-products representation of f is

$$f = \bar{x}_1 \bar{x}_3 + \bar{x}_2 x_3 x_4 + x_1 x_2 x_3 + x_2 x_3 \bar{x}_4 \qquad (4.3)$$

All of the product terms in this representation are essential prime implicants. The boxed area on the left part of Table 4.1, for example, corresponds to the first one, $\bar{x}_1 \bar{x}_3$. It is instructive to work out the others.

Table 4.1   *A two-valued Karnaugh map and a corresponding many-valued connective*

	f	00	01	11	10
	00	1	1	0	0
$x_3 x_4$	01	1	1	0	0
	11	1	0	1	1
	10	0	1	1	0

with $x_1 x_2$ as the top label.

ϕ	g	00	01	11	10
	00	$\frac{2}{3}$	$\frac{1}{3}$	1	0
	01	$\frac{2}{3}$	$\frac{1}{3}$	0	1
	11	$\frac{1}{3}$	1	$\frac{2}{3}$	$\frac{1}{3}$
	10	0	$\frac{1}{3}$	$\frac{2}{3}$	1

with ψ as the top label.

In FM one is often interested in obtaining a *minimal sum-of-product representation* for a given function. It was proven by Quine (1952) that a minimal SOP representation involves only (essential) prime implicants, which restricts the search space considerably. The general problem is co-NP-complete (the minimal SOP of a tautology is an empty product term), although heuristic approaches exist that perform very well in practice if one is content with nearly minimal forms. It is possible to compute nearly minimal representations of circuits that contain thousands of gates.

When we speak of minimizing we always have in mind that the result is minimal in the sense of our tableau rule definition; in other words:

(T0b) minimize the number of product terms; then

(T3) minimize the number of literals in the product terms; in these

(T4) maximize the sets occurring in the literal operators.

This kind of minimization is very common in FM, although there are also algorithms for other criteria.

Now look at the truth table of the four-valued connective g which is drawn on the right in Figure 4.1 (we do not claim that it has a natural interpretation; it is merely a technical example). Assume that we want to determine the rule for $\{\frac{1}{3}, \frac{2}{3}\} g(\phi, \psi)$. We need to find a minimal cover for the entries contained in $\{\frac{1}{3}, \frac{2}{3}\} g(\phi, \psi)$. But these correspond exactly to the ON-set of the two-valued four-place function f on the left. Moreover, the minimal SOP representation of f is just what we want and we can read the conclusion of the rule directly from it, using the following mapping between product terms in the SOP and extensions in the tableau rule.

The first step is the *completion* of the product terms. Whenever a variable x_i does not occur in a product term, add $x_i^{\mathbb{B}}$ to it, where $\mathbb{B} = \mathbf{2} = \{0, 1\}$.

The next step is to map the truth values that the subformulas of g can take on to the variables x_i in f. In our case $x_3 x_4$ represents the truth values that ϕ may take on and $x_1 x_2$ the truth values that ψ may take on. Let $m(\phi) = \{3, 4\}, m(\psi) = \{1, 2\}$ be this mapping.

Then the extension corresponding to a completed product term p is

$\mathsf{S}_\phi \, \phi \circ \mathsf{S}_\psi \, \psi$, where

$$\mathsf{S}_\theta = \{\mathtt{bin2rat}(\vec{x}) | \ \vec{x} \text{ in ON-set of the restriction of } p \text{ to indices in } m(\theta)\}$$

and $\mathtt{bin2rat}$ is a function that converts a minterm read as a binary number to a rational number which represents a truth value in N.

For example, the extension corresponding to the product term $\bar{x}_1 \bar{x}_3$ in the example above is computed as follows.

The completion is $p = x_1^0 x_2^{\mathbb{B}} x_3^0 x_4^{\mathbb{B}}$. The restriction of p to indices in $m(\phi)$ is $x_3^0 x_4^{\mathbb{B}}$. The ON-set of this (with respect to $\{x_3, x_4\}$) is $\{(0, 0), (0, 1)\}$. Hence,

$$\mathsf{S}_\phi = \{\mathtt{bin2rat}((0, 0)), \mathtt{bin2rat}((0, 1))\} = \left\{0, \frac{1}{3}\right\}$$

Similarly, for ψ we obtain

$$S_\psi = \{\texttt{bin2rat}((0,0)), \texttt{bin2rat}((0,1))\} = \left\{0, \frac{1}{3}\right\}$$

A slight complication does occur when the number of truth values n is no power of 2, since we can represent 2^k truth values with k two-valued variables. The solution is simple—just take the smallest k such that $2^k \geq n$ and set all entries that correspond to a greater truth value than n to 0 in the two-valued representation.

We come back to the example with which we opened this section, a combination of sign and connective that gave rise to a non-unique tableau rule. In FM it is well known that the SOP representation of already two-valued functions with literal operators is not unique. In fact, the function f defined above has three minimal SOP representations other than (4.3), one of them being

$$f = \bar{x_1}\bar{x_3} + \bar{x_2}x_3x_4 + x_1x_2x_3 + \bar{x_1}x_2\bar{x_4}$$

Each gives rise to a different tableau rule according to our definition.

Among the criticisms that can be made against our framework as it now stands is the fact that it is not easy to find minimal rules for more complicated logics. Definition 4.4 is not very helpful, as it does not give any indication as to how to find the required homomorphisms. The relationship between FM and tableau rules developed in this section, however, renders the considerable amount of existing work and the very refined algorithms of FM accessible for the purpose of finding many-valued tableau rules. Efficient implementations of standard algorithms are available as Public Domain Software (Yurchak and Butler, 1990).

We summarize the steps that have to be undertaken in order to find a tableau rule for a signed formula S $F(\phi, \psi)$.[*]

1. Create a truth table of the same size as the one for F, whose entries are 1 iff the corresponding entry in the truth table for F is in S and 0 otherwise.

2. If necessary, add rows and columns filled with 0s until a power of 2, say 2^k, is reached.

3. Choose $2k$ two-valued variables, give them indices running from 1 to[†] $2k$, and map the subformulas of $F(\phi, \psi)$ to their indices, as described above.

4. Write down the resulting two-valued $2k$-place function f in non-minimal SOP representation (one product term for each truth table entry).

---

[*]It is not difficult to generalize the method to m-place connectives.
[†]Running to mk for m-place connectives.

5. Compute a (nearly) minimal SOP representation of f with any FM algorithm.

6. Extract the tableau rule from this representation as described above.

Obviously, all steps can be easily automated.

We are now in a position where we can see that Proposition 4.5 holds: if we are given the homomorphisms $g_1, \ldots, g_r$ we compute a SOP representation of the function called f above. Minimization yields a tableau rule from which the required homomorphisms $h_1, \ldots, h_q$ are easily extracted.

5

UNIFORM NOTATION REGAINED: REGULAR LOGICS

SLY. ... *Comes there any more of it?*
PAGE. *My lord, 'tis but begun.*
SLY. *'Tis a very excellent piece of work, madam lady. Would 'twere done!*
— William Shakespeare, *The Taming of the Shrew*

Two of the questions that were raised at the end of Chapter 3 are still open, namely the problem of introducing satisfactory quantifier rules, and the classification problem for many-valued tableau rules. The sets-as-signs notion from the preceding chapter, while answering the other two questions, also posed some new problems, namely the theoretically vast number of signs and the non-trivial computation of rules. The last problem has been alleviated in Section 4.5, but the need for FM tools seems to be unjustified for relatively simple logics.

We will address all these issues in the present chapter. Since the quantifier problem turns out to be related to the others we solve these first, before turning to first-order logics in Section 5.4.

5.1 Primary multiple-valued connectives

A good starting point for the uniform notation problem in many-valued logics is to review the situation in classical logic. Not all classical connectives are associated with an α- or β-rule, which becomes quite obvious by the observation that there are $2^{2^2} = 16$ different binary two-place connectives, eight different α-rules, and eight different β-rules. One needs an α- *and* a β-rule to characterize a connective, consequently there can be at most eight different connectives that are characterized by α- and β-rules. These connectives have been called **primary connectives** by Smullyan (1968) and we have stated their truth tables in Table 5.1 for convenience. A non-primary connective would be, for example, equivalence $\equiv$.

A class of many-valued connectives that play a similar rôle (namely having uniform notation style tableau systems) in many-valued logics as do primary connectives in classical logic would give a class of logics with tableau systems that are close to Smullyan's. So we must characterize primary connectives independently of the number of truth values in order to generalize them.

Table 5.1 *Primary propositional connectives*

$\vee$	0	1
0	0	1
1	1	1

$\downarrow$	0	1
0	1	0
1	0	0

$\wedge$	0	1
0	0	0
1	0	1

$\uparrow$	0	1
0	1	1
1	1	0

$\supset$	0	1
0	1	1
1	0	1

$\not\supset$	0	1
0	0	0
1	1	0

$\subset$	0	1
0	1	0
1	1	1

$\not\subset$	0	1
0	0	1
1	0	0

Definition 5.1. (Conjugate truth value) *Let N be a set of truth values. We define a unary postfix function* $*: N \to N$ *by* $i^* = 1 - i$. *If $S \subseteq N$ we let $S^* = \{i^* \mid i \in S\}$. We call i^* the* **conjugate truth value** *of i.*

Note that

$$v(\neg\phi) = v^*(\phi).$$

We begin with a many-valued version of primary connectives defined with the help of conjugation.

Definition 5.2. ($\mathcal{L}^n_{\mathrm{Sm}}$) *Let $\mathcal{L}^n_{\mathrm{Sm}}$ be the family of n-valued propositional logics whose languages are defined by $(L^n_{Sm}, \neg, \vee, \downarrow, \wedge, \uparrow, \supset, \not\supset, \subset, \not\subset)$ with similarity type $\langle 1, 2, 2, 2, 2, 2, 2, 2, 2 \rangle$, with designated truth values $D = \{\frac{n-d}{n-1}, \ldots, 1\}$, and operator semantics specified as follows:*

$$
\begin{aligned}
v(\neg\phi) &= v^*(\phi) \\
v(\phi \vee \psi) &= max\{v(\phi), v(\psi)\} \\
v(\phi \downarrow \psi) &= min\{v^*(\phi), v^*(\psi)\} \\
v(\phi \wedge \psi) &= min\{v(\phi), v(\psi)\} \\
v(\phi \uparrow \psi) &= max\{v^*(\phi), v^*(\psi)\} \\
v(\phi \supset \psi) &= max\{v^*(\phi), v(\psi)\} \\
v(\phi \not\supset \psi) &= min\{v(\phi), v^*(\psi)\} \\
v(\phi \subset \psi) &= max\{v(\phi), v^*(\psi)\} \\
v(\phi \not\subset \psi) &= min\{v^*(\phi), v(\psi)\}
\end{aligned}
$$

Some of these operators we know already from Definition 2.28. The others are reasonable generalizations along the same lines. If we set $N = \{0,1\}, D = \{1\}$ we obtain precisely the classical primary connectives.

The next problem we must face is how to choose an appropriate set of signs for these logics. If we admit arbitrary signs we know from Figure 4.3 (see, for example, the rule for $\{\frac{1}{2}\}$ $(\phi \vee \psi)$) that we have no chance of

Table 5.2 α- and β-formulas and their components for $\mathcal{L}^n_{\mathrm{Sm}}$

α	α_1	α_2
$\mathsf{D}\,\neg\phi$	$\mathsf{D}^*\,\phi$	$\mathsf{D}^*\,\phi$
$\mathsf{D}\,\phi\wedge\psi$	$\mathsf{D}\,\phi$	$\mathsf{D}\,\psi$
$\mathsf{D}\,\phi\downarrow\psi$	$\mathsf{D}^*\,\phi$	$\mathsf{D}^*\,\psi$
$\mathsf{D}\,\phi\not\supset\psi$	$\mathsf{D}\,\phi$	$\mathsf{D}^*\,\psi$
$\mathsf{D}\,\phi\not\subset\psi$	$\mathsf{D}^*\,\phi$	$\mathsf{D}\,\psi$
$\bar{\mathsf{D}}\,\neg\phi$	$\bar{\mathsf{D}}^*\,\phi$	$\bar{\mathsf{D}}^*\,\phi$
$\bar{\mathsf{D}}\,\phi\vee\psi$	$\bar{\mathsf{D}}\,\phi$	$\bar{\mathsf{D}}\,\psi$
$\bar{\mathsf{D}}\,\phi\uparrow\psi$	$\bar{\mathsf{D}}^*\,\phi$	$\bar{\mathsf{D}}^*\,\psi$
$\bar{\mathsf{D}}\,\phi\subset\psi$	$\bar{\mathsf{D}}\,\phi$	$\bar{\mathsf{D}}^*\,\psi$
$\bar{\mathsf{D}}\,\phi\supset\psi$	$\bar{\mathsf{D}}^*\,\phi$	$\bar{\mathsf{D}}\,\psi$
$\mathsf{D}^*\,\neg\phi$	$\mathsf{D}\,\phi$	$\mathsf{D}\,\phi$
$\mathsf{D}^*\,\phi\vee\psi$	$\mathsf{D}^*\,\phi$	$\mathsf{D}^*\,\psi$
$\mathsf{D}^*\,\phi\uparrow\psi$	$\mathsf{D}\,\phi$	$\mathsf{D}\,\psi$
$\mathsf{D}^*\,\phi\subset\psi$	$\mathsf{D}^*\,\phi$	$\mathsf{D}\,\psi$
$\mathsf{D}^*\,\phi\supset\psi$	$\mathsf{D}\,\phi$	$\mathsf{D}^*\,\psi$
$\bar{\mathsf{D}}^*\,\neg\phi$	$\bar{\mathsf{D}}\,\phi$	$\bar{\mathsf{D}}\,\phi$
$\bar{\mathsf{D}}^*\,\phi\wedge\psi$	$\bar{\mathsf{D}}^*\,\phi$	$\bar{\mathsf{D}}^*\,\psi$
$\bar{\mathsf{D}}^*\,\phi\downarrow\psi$	$\bar{\mathsf{D}}\,\phi$	$\bar{\mathsf{D}}\,\psi$
$\bar{\mathsf{D}}^*\,\phi\not\supset\psi$	$\bar{\mathsf{D}}^*\,\phi$	$\bar{\mathsf{D}}\,\psi$
$\bar{\mathsf{D}}^*\,\phi\not\subset\psi$	$\bar{\mathsf{D}}\,\phi$	$\bar{\mathsf{D}}^*\,\psi$

β	β_1	β_2
$\mathsf{D}\,\phi\vee\psi$	$\mathsf{D}\,\phi$	$\mathsf{D}\,\psi$
$\mathsf{D}\,\phi\uparrow\psi$	$\mathsf{D}^*\,\phi$	$\mathsf{D}^*\,\psi$
$\mathsf{D}\,\phi\subset\psi$	$\mathsf{D}\,\phi$	$\mathsf{D}^*\,\psi$
$\mathsf{D}\,\phi\supset\psi$	$\mathsf{D}^*\,\phi$	$\mathsf{D}\,\psi$
$\bar{\mathsf{D}}\,\phi\wedge\psi$	$\bar{\mathsf{D}}\,\phi$	$\bar{\mathsf{D}}\,\psi$
$\bar{\mathsf{D}}\,\phi\downarrow\psi$	$\bar{\mathsf{D}}^*\,\phi$	$\bar{\mathsf{D}}^*\,\psi$
$\bar{\mathsf{D}}\,\phi\not\supset\psi$	$\bar{\mathsf{D}}\,\phi$	$\bar{\mathsf{D}}^*\,\psi$
$\bar{\mathsf{D}}\,\phi\not\subset\psi$	$\bar{\mathsf{D}}^*\,\phi$	$\bar{\mathsf{D}}\,\psi$
$\mathsf{D}^*\,\phi\wedge\psi$	$\mathsf{D}^*\,\phi$	$\mathsf{D}^*\,\psi$
$\mathsf{D}^*\,\phi\downarrow\psi$	$\mathsf{D}\,\phi$	$\mathsf{D}\,\psi$
$\mathsf{D}^*\,\phi\not\supset\psi$	$\mathsf{D}^*\,\phi$	$\mathsf{D}\,\psi$
$\mathsf{D}^*\,\phi\not\subset\psi$	$\mathsf{D}\,\phi$	$\mathsf{D}^*\,\psi$
$\bar{\mathsf{D}}^*\,\phi\vee\psi$	$\bar{\mathsf{D}}^*\,\phi$	$\bar{\mathsf{D}}^*\,\psi$
$\bar{\mathsf{D}}^*\,\phi\uparrow\psi$	$\bar{\mathsf{D}}\,\phi$	$\bar{\mathsf{D}}\,\psi$
$\bar{\mathsf{D}}^*\,\phi\subset\psi$	$\bar{\mathsf{D}}^*\,\phi$	$\bar{\mathsf{D}}\,\psi$
$\bar{\mathsf{D}}^*\,\phi\supset\psi$	$\bar{\mathsf{D}}\,\phi$	$\bar{\mathsf{D}}^*\,\psi$

achieving a uniform notation style system. On the other hand, do we really want to know whether $\{\frac{1}{2}\}\,(\phi\vee\psi)$? A much more common query is the one for either $\mathsf{D}\,\phi$ or $\mathsf{N}-\mathsf{D}\,\phi$, that is, whether or not ϕ has a designated truth value. Let us abbreviate $\mathsf{N}-\mathsf{D}$ to $\bar{\mathsf{D}}$.

Calculation of the tableau rules for the connectives in $\mathcal{L}^n_{\mathrm{Sm}}$ and for the signs D and $\bar{\mathsf{D}}$ shows that in the conclusions of the resulting tableau rules no signs other than $\mathsf{D},\bar{\mathsf{D}},\mathsf{D}^*,\bar{\mathsf{D}}^*$ occur. The same holds in the conclusions of tableau rules for D^* and $\bar{\mathsf{D}}^*$. Thus it is possible to fix the set of signs for $\mathcal{L}^n_{\mathrm{Sm}}$ as

$$\mathsf{S}_{\mathcal{L}^n_{\mathrm{Sm}}} = \{\mathsf{D},\bar{\mathsf{D}},\mathsf{D}^*,\bar{\mathsf{D}}^*\}$$

and obtain a sound and complete system from it. Moreover, it turns out that all rules obey the α- or β-schema and we can present the whole tableau system as in Table 5.2, together with the usual α- and β-schemata. For convenience we treat negated formulas as α-formulas where $\alpha_1 = \alpha_2$.

So we have constructed a uniform notation style tableau system for testing the validity of formulas in any logic that contains at most negation plus the generalized versions of the classical primary operators and where D is of the form $\{\frac{n-d}{n-1},\ldots,1\}$. To test the validity of a formula, say ϕ, all we have to do is to construct a single closed tableau with root $\bar{\mathsf{D}}\,\phi$.

Automated tableau construction is possible with any tableau-based theorem prover that can be modified to accommodate four instead of two signs. On the other hand, the well-developed classical proofs of tableau completeness and soundness (Fitting, 1990b) can be carried over from classical logic with only minor modifications.[*] This is particularly useful in the case of implementation oriented tableau proof systems, where the completeness proof tends to be awkward.

The reason why the conjugates of D and $\bar{\mathsf{D}}$ must also be present is that, in general, the equality $\bar{\mathsf{D}} = \mathsf{D}^*$ does not hold. The signs D and $\bar{\mathsf{D}}$ represent designated and non-designated truth values, whereas the signed formula $\mathsf{D}^*\,\phi$ is equivalent to $\mathsf{D}\,\neg\phi$. In classical logic, negation and non-designatedness coincide. This is the reason why logically equivalent signed and non-signed versions of classical tableau systems can be formulated without extending the set of signs or the set of logical connectives.

Our resulting logics are not very exciting, though. Since the expansion rules are the same for all n, a different logic can only arise when branches are closed. It is easy to see that all logics in $\mathcal{L}^n_{\mathrm{Sm}}$ fall into only three classes, namely when one of

$$\bar{\mathsf{D}} \;=\; \mathsf{D}^* \qquad\qquad (5.1)$$

$$\bar{\mathsf{D}} \;\subsetneq\; \mathsf{D}^* \qquad\qquad (5.2)$$

$$\bar{\mathsf{D}} \;\supsetneq\; \mathsf{D}^* \qquad\qquad (5.3)$$

is satisfied. Within each class, all logics have essentially the same set of tautologies and the same consequence relation.[†] A moment's reflection shows that the following cases can actually occur.

If (5.1) holds, then, also $\bar{\mathsf{D}}^* = \mathsf{D}$. If we substitute

$$\mathsf{T} \quad \text{for} \quad \bar{\mathsf{D}}^* = \mathsf{D}$$

$$\mathsf{F} \quad \text{for} \quad \bar{\mathsf{D}} = \mathsf{D}^*$$

the component formula Table 5.2 collapses into the classical one. Moreover, since

$$S_1 \cap S_2 = \emptyset \text{ iff } S_1 \in \{\bar{\mathsf{D}}, \mathsf{D}^*\} = \{\mathsf{F}\,\} \text{ and } S_2 \in \{\bar{\mathsf{D}}^*, \mathsf{D}\} = \{\mathsf{T}\,\}$$

or vice versa, the resulting proof system amounts exactly to the classical tableau system for the primary connectives. By strong soundness of classical tableaux we have that $\mathcal{L}^n_{\mathrm{Sm}}$ is classical propositional logic when $\bar{\mathsf{D}} = \mathsf{D}^*$.

---

[*] Of course the results also follow from the preceding chapter; the completeness of the set of signs $\mathbf{S}_{\mathcal{L}^n_{\mathrm{Sm}}}$ with respect to $\mathcal{L}^n_{\mathrm{Sm}}$ is easily checked.

[†] One can obtain some other logics when fragments are considered.

The remaining cases occur only when $n \geq 3$. If (5.3) holds we reason in the same way as above that there is only a single tableau system, namely by noting that the closure is independent of the choice of $\bar{\mathsf{D}}, \mathsf{D}^*, \bar{\mathsf{D}}^*, \mathsf{D}$ under that proviso. Consequently, the set

$$DL = \{\phi| \ \bar{\mathsf{D}} \ \phi \text{ has a closed tableau}\}$$

is also independent of D. Therefore, we can take any logic from $\mathcal{L}^n_{\mathrm{Sm}}$ as a model. It turns out that the restriction to the connectives $\neg, \wedge, \vee, \supset$ with $n = 3, D = \{1\}$ yields three-valued strong Kleene logic, see Section 2.3.1. Since all primary connectives are definable by only $\neg$ and $\vee$, DL is essentially the set of tautologies of $\mathcal{L}^n_{\mathrm{SKL}}$.

Finally, if (5.2) holds we obtain a kind of dual to $\mathcal{L}^n_{\mathrm{SKL}}$ which can be characterized with the help of

$$D^*L = \{\phi| \ \mathsf{D}^* \ \phi \text{ has a closed tableau}\}$$

by the equality $D^*L_{(5.2)} = DL_{(5.3)}$ (under the respective provisos indicated by subscripts). A dual result holds as well.

This is a little disappointing; however, we have only considered a very special class of signs and connectives. There are more general possibilities for combining signs and rules so as to preserve the validity of the α- and β- rule schemata. Although a characterization of all these combinations is outside the scope of this book, in the following section we shall gain access to a substantial class of logics arising in this way.

5.2 Regular logics

The key is to observe a certain regularity that can be found in the truth tables of many multiple-valued functions and, in particular, in those of the primary connectives.

FIG. 5.1. Shapes of the truth table entries to be covered in truth tables of primary connectives.

The shape of the truth table entries to be covered in truth tables of primary connectives combined with signs from $\mathbf{S}_{\mathcal{L}^n_{\mathrm{Sm}}}$ is always one of the shaded areas in Figure 5.1, where the corner that the pattern is starting from is arbitrary. The main reasons for this very simple shape are:

- We only have signs of the form $\{0, \ldots, i\}$ and $\{j, \ldots, 1\}$.
- The truth values corresponding to the truth table entries are monotonically increasing or decreasing, starting from one corner.

We express these constraints in a formal definition.

Definition 5.3. (Circle, corner set) *We define a metric* * on N^k by*

$$d(\vec{x}, \vec{y}) = \max_{1 \leq i \leq k} |x_i - y_i|$$

*for $\vec{x}, \vec{y} \in N^k$. For $r \in N$ we define the **circle** in N^k with center $\vec{x}$ and radius r as the set*

$$c_{\vec{x}, r} = \{\vec{y}|\ \vec{y} \in N^k \text{ and } d(\vec{x}, \vec{y}) = r\}$$

*We define the **corner set** of N^k as*

$$I^k = \{\vec{x}|\ x_i \in \{0,1\}, 1 \leq i \leq k\} = \{0,1\}^k$$

Definition 5.4. (Regular operator) *A many-valued k-place connective f is called **regular** iff there is a $\vec{x} \in I^k$ such that*

1. *for all $r \in N$ $f(c_{\vec{x},r})$ is a singleton, say $\{x_r\}$, and*
2. *the sequence $x_0, \ldots, x_1$ is monotonic (either increasing or decreasing) wrt the natural order on N.*

*We call $\vec{x}$ the **starting point** of f.*

The entries in truth tables of regular operators are ordered starting from some corner with the lowest (highest) truth value that the operator takes on. On each circle, with the starting point as its centre, all entries contain the same truth value. These truth values are monotonically increasing (decreasing) with the radius of the circles. Figure 5.2 shows a typical pattern, while Table 5.3 shows an example of a regular operator. Also, the operators $\{\wedge, \vee, \neg, J_0, J_1, \sim\}$ of $\mathcal{L}^n_{\mathrm{M}+}$ are regular.

We define parameters which are sufficient to describe all generalized primary connectives. Let I stand for the identity function on N. Recall that $*$ was the conjugate function on N. Furthermore, we abbreviate a truth value vector with identical entries $(i, i, \ldots, i)$ to $\vec{i}$.

Definition 5.5. (Direction, orientation) *Let f be a regular operator and $\vec{x}$ its starting point. Then $\vec{\delta}(f) = (\delta(x_1), \ldots, \delta(x_k))$ maps regular*

---

*We could call this metric Chessboard Metric, since the distance gives the minimal number of fields that have to be passed between two points when diagonal moves are allowed.

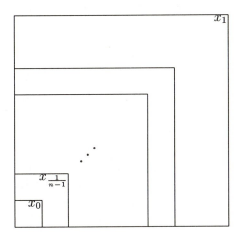

FIG. 5.2. Typical pattern of regular operator.

Table 5.3 *Truth table for operator $\odot$. The marked entries constitute a circle with radius $\frac{2}{3}$ and centre $(1,0)$*

$\odot$	0	$\frac{1}{3}$	$\frac{2}{3}$	1
0	0	0	0	0
$\frac{1}{3}$	$\boxed{\frac{1}{3}}$	$\boxed{\frac{1}{3}}$	$\boxed{\frac{1}{3}}$	0
$\frac{2}{3}$	1	1	$\boxed{\frac{1}{3}}$	0
1	1	1	$\boxed{\frac{1}{3}}$	0

connectives into $\{^*, I\}^k$, is defined by $\delta(0) := I, \delta(1) := {}^*$ and is called the **orientation** of f, while

$$\epsilon(f) = \begin{cases} I & \text{if } (f(c_{\vec{x},i}))_{0 \leq i \leq 1} \text{ is monotonically increasing} \\ {}^* & \text{if } (f(c_{\vec{x},i}))_{0 \leq i \leq 1} \text{ is monotonically decreasing} \end{cases}$$

is called the **direction** of f. Write $\delta_i(f)$ for the ith component of $\vec{\delta}(f)$.

Definition 5.6. (Threshold) *If $\delta(f) = (I, I)$ and $\epsilon(f) = I$ we define* **thresholds** *for an operator f and a truth value $i \in N$ as:*

$$t_{f,i} = \min\{j \mid f(\vec{j}) \geq i\} \tag{5.4}$$

$$\bar{t}_{f,i} = \max\{j \mid f(\vec{j}) \leq i\} \tag{5.5}$$

Example 5.7. *Let $\odot$ be defined as in Table 5.3. Then $\delta_1(\odot) = {}^*, \delta_2(\odot) = I, \epsilon(\odot) = {}^*$. Consider $\vee$ for $n = 3$: $\delta_1(\vee) = \delta_2(\vee) = \epsilon(\odot) = I$ and $t_{\vee,1} = 1, \bar{t}_{\vee,1} = 1$.*

If we set $k = 2$ and $(f(c_{\vec{x},i}))_{0 \leq i \leq 1} = (i)_{0 \leq i \leq 1}$ or $(f(c_{\vec{x},i}))_{0 \leq i \leq 1} = (i^*)_{0 \leq i \leq 1}$ the eight possible combinations of $\vec{\delta} = (\delta_1, \delta_2)$ and ϵ yield exactly the eight many-valued primary connectives for each n and the eight classical primary connectives for $n = 2$ in particular.

By $\vec{\delta}$ and the sequence $(f(c_{\vec{x},i}))_{0 \leq i \leq 1}$ a regular operator is completely determined.

From now on we write δ instead of $\vec{\delta}$ and we concentrate on binary operators in order to simplify the presentation. We remark that all considerations carry over to arbitrary k-place operators in a straightforward manner.

Our next goal is a tableau proof system for regular operators with a sufficient number of rules and signs for complete handling of queries of the form $\{0, \ldots, i - \frac{1}{n-1}\}$ and $\{i + \frac{i}{n-1}, \ldots, 1\}$ (in particular, $D\,\phi$ and $\bar{D}\,\phi$). From Figures 5.3 and 5.4 it is obvious that we need only consider signs of the form $\{0, \ldots, t_{f,i} - \frac{1}{n-1}\}$ and $\{\bar{t}_{f,i} + \frac{1}{n-1}, \ldots, 1\}$.

As a shorthand for the signs involved we write the following definition.*

Definition 5.8.

$$\boxed{<i} = \{0, \ldots, \frac{i-1}{n-1}\} = [0, i) \cap N$$

$$\boxed{>i} = \{\frac{i+1}{n-1}, \ldots, 1\} = (i, 1] \cap N$$

Sets of truth values of the form $\boxed{>i}$, $\boxed{<i}$ turn out to be useful in several contexts within many-valued logics. In Post algebras (Epstein, 1960) they correspond exactly to the set of truth values characterized by the important D_i operators, see also Section 8.1.4.3. The lattice theoretic notions of *upset* and *downset* boil down to $\boxed{>i}$, $\boxed{<i}$ in the case of linear orders, see Section 5.5 for a further discussion.

If $\delta(f) = (I, I), \epsilon(f) = I$, the following obviously holds (compare Figures 5.3 and 5.4):

---

*If $i = 0$ (respectively, $i = 1$) we define these sets to be empty.

$f(\phi, \psi)$

0

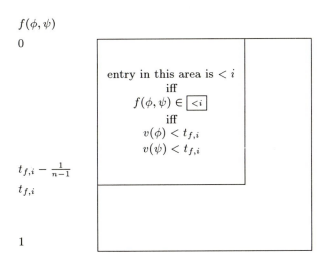

$t_{f,i} - \frac{1}{n-1}$

$t_{f,i}$

1

FIG. 5.3. Determining the tableau rules for $\boxed{<i}\, f(\phi, \psi)$.

$$f(\phi,\psi) \in \boxed{<i} \quad \text{iff} \quad v(\phi) \in \boxed{<t_{f,i}} \text{ and } v(\psi) \in \boxed{<t_{f,i}}$$

$$f(\phi,\psi) \in \boxed{>i} \quad \text{iff} \quad v(\phi) \in \boxed{>\bar{t}_{f,i}} \text{ or } v(\psi) \in \boxed{>\bar{t}_{f,i}}$$

Consequently, provided that in the premisses of our rules only signs of the form $\boxed{>i}$ and $\boxed{<i}$ occur, we also have only signs of the same type in the conclusions. This motivates the following definition.

Definition 5.9. (Regular logic) *A* **regular logic** *is any logic $\mathcal{L}$ containing only regular connectives and the set of signs is fixed as*

$$\mathbf{S}_{\mathcal{L}} = \{\boxed{>i} \mid i \in N, i \neq 1\} \cup \{\boxed{<i} \mid i \in N, i \neq 0\}.$$

Moreover, we call a set of signs of this shape a **regular set of signs**.

We note that in *any* tableau system with a regular set of signs we can restrict the check for closure to pairs of signed formulas (in Definition 4.7 we had to account for tupels). It is easy to see why this is sufficient: assume that we have three identical formulas, but with different signs. Then either two of them are of the form $\boxed{>i_1}\,\phi$, $\boxed{>i_2}\,\phi$ or two are of the form $\boxed{<i_3}\,\phi$, $\boxed{<i_4}\,\phi$. In both cases one formula is subsumed.

Theorem 5.10. *Let $\mathcal{L}$ be a regular logic such that for each connective f $\delta(f) = (I, I), \epsilon(f) = I$. Then the tableau system defined by the following α and β component rules for each f is sound and complete.*

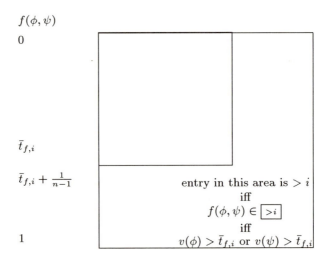

FIG. 5.4. Determining the tableau rules for $\boxed{>i}\, f(\phi, \psi)$.

α	α_1	α_2
$\boxed{<i}\, f(\phi, \psi)$	$\boxed{<t_{f,i}}\, \phi$	$\boxed{<t_{f,i}}\, \psi$

β	β_1	β_2
$\boxed{>i}\, f(\phi, \psi)$	$\boxed{>\bar{t}_{f,i}}\, \phi$	$\boxed{>\bar{t}_{f,i}}\, \psi$

Proof. Follows directly from the above observations and Theorems 4.14 and 4.21, since a regular set of signs is complete with respect to regular connectives.

Before we can deal with the other cases, where one or more of the parameters is equal to *, we must state some basic properties of conjugation.

Lemma 5.11.
1. $\left(\boxed{<i}\right)^* = \boxed{>i^*}$ for all $i \in N$.
2. $\left(\boxed{>i}\right)^* = \boxed{<i^*}$ for all $i \in N$.
3. $(i^*)^* = i$ for all $i \in N$.
4. $(S^*)^* = S$ for all $S \subseteq N$.

Proof. Follows almost immediately from the definitions.

To change the rules for proper handling of values other than I of δ_i, observe that $\delta_1 = *$ indicates that the truth table of f is flipped around its horizontal axis. To compute the threshold function properly we must count from upside down in the first argument of f or, in other words, we must conjugate the first argument before computing the threshold; however, the result has to be conjugated again. The same considerations also apply, of course, to the second argument of f and we arrive at the following updated definitions for the threshold functions and component rules.

Definition 5.12. (Threshold, tableau rules $\delta \in \{I, *\}, \epsilon = I$)

$$t_{f,i} \;\; = \;\; \min\{j \mid f(\vec{j}^{\,\delta(f)}) \geq i\} \tag{5.6}$$

$$\bar{t}_{f,i} \;\; = \;\; \max\{j \mid f(\vec{j}^{\,\delta(f)}) \leq i\} \tag{5.7}$$

$\vec{j}^{\;\delta(f)}$ is to be read as $(j_1^{\cdot\delta_1(f)}, j_2^{\cdot\delta_2(f)})$. The component rules for α- and β-formulas are given in the following table:

α	α_1	α_2
$\boxed{<i}\, f(\phi, \psi)$	$\left(\boxed{<t_{f,i}}\right)^{\delta_1(f)} \phi$	$\left(\boxed{<t_{f,i}}\right)^{\delta_2(f)} \psi$

β	β_1	β_2
$\boxed{>i}\, f(\phi, \psi)$	$\left(\boxed{>\bar{t}_{f,i}}\right)^{\delta_1(f)} \phi$	$\left(\boxed{>\bar{t}_{f,i}}\right)^{\delta_2(f)} \psi$

The next theorem follows immediately from what has been said above.

Theorem 5.13. *Let $\mathcal{L}$ be a regular logic with the restriction that for all connectives f $\epsilon(f) = I$. Then the tableau system given by the α and β component rules in Definition 5.12 is sound and complete.*

Example 5.14. *Consider the three-valued weak implication from Definition 5.2. Assume that we want to compute the rule for $\{\frac{1}{2}, 1\}$ $(\phi \supset \psi)$:*

- *$\{\frac{1}{2}, 1\} = \boxed{>0}$, so our component rules tell us that we have a β-type rule with extensions $\left(\boxed{>\bar{t}_{\supset,0}}\right)^{\delta_1(\supset)} \phi$ and $\left(\boxed{>\bar{t}_{\supset,0}}\right)^{\delta_2(\supset)} \psi$.*
- *Since $\delta(\supset) = (*, I)$, we have $\bar{t}_{\supset,0} = \max\{j \mid v(j^* \supset j) \leq 0\} = 0$, which yields extensions $\left(\boxed{>0}\right)^* \phi$ and $\boxed{>0}\, \psi$.*
- *$\left(\boxed{>0}\right)^* = \boxed{<0^*} = \boxed{<1} = \{0, \frac{1}{2}\}$ and $\boxed{>0} = \{\frac{1}{2}, 1\}$ finally give the rule*

$$\frac{\{\frac{1}{2}, 1\}\,(\phi \supset \psi)}{\{0, \frac{1}{2}\}\, \phi \mid \{\frac{1}{2}, 1\}\, \psi}$$

We can treat strong negation as a special case of the above stated rules. If we note that $\alpha_1(\neg) = \alpha_2(\neg)$, make the convention that $\beta_1(\neg) = \beta_2(\neg)$,

Table 5.4   *Negation rules in regular logics*

$$\frac{\boxed{<i}\,\neg\phi}{\boxed{>i^*}\,\phi} \qquad\qquad \frac{\boxed{>i}\,\neg\phi}{\boxed{<i^*}\,\phi}$$

note that $\delta_i(\neg) = {}^*$ and $t_{\neg,i} = \bar{t}_{\neg,i} = i$ for all i, treat extensions as sets, and omit identical extensions, we arrive at the negation rules stated in Table 5.4. The unary regular connectives $\sim$ and J_0, J_1 can be handled similarly.

We still have to treat the case when $\epsilon(f) = {}^*$. Taking into account this possibility in the component rules and the threshold functions leads to the following definitions.

Definition 5.15. (Conjugate operator) *Let f be a regular operator. We define the* **conjugate operator** *of f, denoted by f^* as*

$$f^*(i,j) := \neg(f(\neg i, \neg j)) = (f(i^*, j^*))^*$$

It easy to prove that the conjugate of a regular operator is again regular.

Definition 5.16. (Threshold, tableau rules—general case)

$$t_{f,i} = \min\{j \mid v(f(\vec{j}^{\,\delta(f)})) \geq^{\epsilon(f)} i\} \qquad (5.8)$$
$$\bar{t}_{f,i} = \max\{j \mid v(f(\vec{j}^{\,\delta(f)})) \leq^{\epsilon(f)} i\} \qquad (5.9)$$

where $\leq^* := \geq$ and $\geq^* := \leq$.

α		α_1		α_2	
$\left(\boxed{<i^*}\right)^{\epsilon(f)}\ f(\phi,\psi)$		$\left(\boxed{<t_{f^{\epsilon(f)},i}^*}\right)^{\delta_1(f)}\ \phi$		$\left(\boxed{<t_{f^{\epsilon(f)},i}^*}\right)^{\delta_2(f)}\ \psi$	

β		β_1		β_2	
$\left(\boxed{>i^*}\right)^{\epsilon(f)}\ f(\phi,\psi)$		$\left(\boxed{>\bar{t}_{f^{\epsilon(f)},i}^*}\right)^{\delta_1(f)}\ \phi$		$\left(\boxed{>\bar{t}_{f^{\epsilon(f)},i}^*}\right)^{\delta_2(f)}\ \psi$	

Example 5.17. *Let $\odot$ be defined as in Table 5.3. $t_{\odot,\frac{2}{3}} = \frac{2}{3}$, $\bar{t}_{\odot,\frac{2}{3}} = \frac{2}{3}$.*

Lemma 5.18. *For all $i, j \in N$, regular operators f:*

1. $f(i,j) = \neg(f^*(\neg i, \neg j)) = (f^*(i^*, j^*))^*$.
2. $\delta_i(f^*) = \delta_i^{\,*}(f)$.
3. $\epsilon(f^*) = \epsilon^*(f)$.

Proof.   Immediate consequence of the definitions.

Theorem 5.19. *Let $\mathcal{L}$ be any regular logic. Then the tableau system given by the α and β component rules in Definition 5.16 is sound and complete.*

Proof. The proof is by reduction to the case $\epsilon(f) = I$ (Theorem 5.13).

If $\epsilon(f) = I$, substitute i^* for i and apply Lemma 5.11 to arrive at the rules for $\epsilon(f) = I$.

If $\epsilon(f) = *$, from Lemma 5.18(3) we conclude that $\epsilon(f^*) = I$.

Starting with the premise of the β component rule we proceed as follows, using Lemma 5.18(1), Lemma 5.11, and the negation rules:

$$\left(\boxed{>i^*}\right)^{\epsilon(f)} f(\phi, \psi) \;\equiv\; \left(\boxed{>i^*}\right)^{*} \neg(f^*(\neg\phi, \neg\psi))$$

$$\equiv\; \boxed{<i}\,\neg(f^*(\neg\phi, \neg\psi))$$

$$\equiv\; \boxed{>i^*}\,f^*(\neg\phi, \neg\psi)$$

Now we can use the β-rule for $\epsilon = I$, since $\epsilon(f^*) = I$. This gives us the extensions

$$\left(\boxed{>\bar{t}_{f^*,i^*}}\right)^{\delta_1(f^*)} \neg\phi \qquad \text{and} \qquad \left(\boxed{>\bar{t}_{f^*,i^*}}\right)^{\delta_2(f^*)} \neg\psi$$

From here the theorem follows by Lemma 5.18(2), Lemma 5.11, and the negation rules again.

The proof for the α case is similar.

Before giving another example we want to simplify the rules a little. To this end we need the following lemma:

Lemma 5.20. *For all $i \in N$ and all regular operators f:*

1. $t_{f^*,i^*} = t_{f,i}$.
2. $\bar{t}_{f^*,i^*} = \bar{t}_{f,i}$.

Proof. We show only (1); (2) is similar.

By definition we have

$$t_{f^*,i^*} \;=\; \min\{j \mid f^*(\vec{\jmath}^{\,\delta(f^*)}) \geq^{\epsilon(f^*)} i^*\}$$

By definition of f^*, Lemma 5.18(2), and Lemma 5.11:

$$=\; \min\{j \mid (f(\vec{\jmath}^{\,\delta(f)}))^* \geq^{\epsilon(f^*)} i^*\}$$

And since $i^* \leq j^*$ iff $i \geq j$:

$$=\; \min\{j \mid f(\vec{\jmath}^{\,\delta(f)}) \leq^{\epsilon(f^*)} i\}$$

Finally, from Lemma 5.18(3) and by definition of $\leq^*$:

$$= \min\{j \mid f(\vec{j}^{\,\delta(f)}) \geq^{\epsilon(f)} i\}$$
$$= t_{f,i}$$

Using Lemma 5.11 and Lemma 5.20 we can simplify the component rules for the case $\epsilon(f) = \ {}^*$ to:

α	α_1	α_2
$\boxed{>i}\, f(\phi, \psi)$	$\left(\boxed{<t_{f,i}}\right)^{\delta_1(f)}\ \phi$	$\left(\boxed{<t_{f,i}}\right)^{\delta_2(f)}\ \psi$

β	β_1	β_2
$\boxed{<i}\, f(\phi, \psi)$	$\left(\boxed{>\bar{t}_{f,i}}\right)^{\delta_1(f)}\ \phi$	$\left(\boxed{>\bar{t}_{f,i}}\right)^{\delta_2(f)}\ \psi$

Combining this with the earlier result for $\epsilon(f) = I$ we arrive at the following final version.

Definition 5.21. (Rules for regular logics, final version)

α	α_1	α_2
$\left(\boxed{<i^{\epsilon(f)}}\right)^{\epsilon(f)}\ f(\phi, \psi)$	$\left(\boxed{<t_{f,i}}\right)^{\delta_1(f)}\ \phi$	$\left(\boxed{<t_{f,i}}\right)^{\delta_2(f)}\ \psi$

β	β_1	β_2
$\left(\boxed{>i^{\epsilon(f)}}\right)^{\epsilon(f)}\ f(\phi, \psi)$	$\left(\boxed{>\bar{t}_{f,i}}\right)^{\delta_1(f)}\ \phi$	$\left(\boxed{>\bar{t}_{f,i}}\right)^{\delta_2(f)}\ \psi$

Corollary. Let $\mathcal{L}$ be any regular logic. Then the tableau system given by the α and β component rules in Definition 5.21 is sound and complete.

We close this section with an example that uses the final version of our system.

Example 5.22. *We illustrate the rules with the operator $\odot$ defined in Table 5.3. Suppose we were interested in the tableau rule for sign $\{1\}$. From* $\{1\} = \boxed{>\frac{2}{3}} = \left(\boxed{<\frac{2}{3}^*}\right)^*$ *and Examples 5.7 and 5.17 we see that the generic α-rule may be instantiated, yielding the extensions:* $\left(\boxed{<\frac{2}{3}}\right)^* \phi = \boxed{>\frac{1}{3}}\, \phi = \{\frac{2}{3}, 1\}\, \phi$ *and* $\boxed{<\frac{2}{3}}\, \psi = \{0, \frac{1}{3}\}\, \psi$.

5.3   On the scope of regular logics

In this section we draw breath and discuss what has been gained so far. The main result of the last section was the discovery that there is a non-trivial class of many-valued functions, called regular operators, which, together

Table 5.5   *Truth table of σ, σ' and $\neg J_1$*

ϕ	$\sigma(\phi)$	$\sigma'(\phi)$	$\neg J_1(\phi)$
0	$\frac{1}{n-1}$	$\frac{1}{n-1}$	1
$\frac{1}{n-1}$	$\frac{2}{n-1}$	$\frac{2}{n-1}$	1
$\ldots$	$\ldots$	$\ldots$	$\ldots$
$\frac{k}{n-1}$	$\frac{k+1}{n-1}$	$\frac{k+1}{n-1}$	1
$\ldots$	$\ldots$	$\ldots$	$\ldots$
$\frac{n-2}{n-1}$	1	1	1
1	0	1	0

with a suitable choice of signs, have uniform notation style tableau systems. Moreover, the rules of these tableau systems can be calculated in a simple and schematic manner. Checking whether a many-valued function is regular is also easy when its truth table is available. One of the immediate consequences is that for regular logics, any automated tableau proving system can be used with some very slight modifications regarding the number of signs and the check for branch closure.

Another question, however, is how important regular logics are and thus how big the achievement actually is. An encouraging point in this direction is the following

Theorem 5.23. (Functional completeness of regular logics)   *For any n there is a regular logic $\mathcal{L}$ with n truth values which is functionally complete.*

Proof.   It is well known that the Post logics $\mathcal{L}_P^n$ from Definition 2.32 are functionally complete for each n. This is proved, for example, in Urquhart (1986). Unfortunately, σ is not a regular operator, although $\vee$ and $\wedge$ are.

On the other hand, σ is easily definable using regular operators alone, and this is sufficient to prove our claim. To see that σ is definable by regular operators, consider the truth tables for σ, σ', and $\neg J_1$ in Table 5.5.

All $\wedge$, σ', and $\neg J_1$ are regular and obviously

$$\sigma(\phi) \cong \sigma'(\phi) \wedge \neg J_1(\phi)$$

holds, so the desired logic is defined by the matrix $(N, \vee, \wedge, \sigma', \neg, J_1)$.

Later (cf. Section 8.1.1, p. 124 f.) we will see that functional completeness alone is necessary (but not sufficient) for a truly general system intended for real applications. The fact that a logic is functionally complete is of limited use when the number of connectives is very restricted and definitions of new operators are bulky.

In our framework we have a broad selection of connectives—any choice of regular operators is allowed. So even if an operator is not regular itself, there is a good chance that it can be defined concisely with regular operators. For example, weak implication is not regular. It can, however, easily be defined by

$$\phi \supset \psi \equiv \sim \phi \vee \psi.$$

Compare this with the definition in $\mathcal{L}^n_{M+}$ (cf. Sections 8.1.1 and 2.3.1):

$$\phi \supset \psi \equiv \left(\neg (J_{\frac{n-d}{n-1}}(\phi) \vee \cdots \vee J_1(\phi)) \right) \vee \psi$$

Another important point is the expressivity of the sign language. Using regular signs we cannot only state whether a formula takes on truth values from the sets $\{0, \ldots, i\}$ and $\{j, \ldots, 1\}$ for any $i, j \in N$, but it is also easy to combine these constraints conjunctively in order to query for arbitrary intervals or for exact truth values. This does not cause too much branching, since one of the formulas $\{0, \ldots, i\} \phi$, $\{j, \ldots, 1\} \phi$ is always of type α.

In other systems such queries must be encoded by more complex formulas. This is not only unnatural, but usually more expensive.

Combinatorial excursus on the number of regular functions

How many regular operators are there for each n? If m is the number of monotonically increasing sequences with length n and members from N, the answer is $8m - 8n + 1$ (each constant function counts only once in each category). An approximation of m can be computed as follows. For each $k \in N$ there is one sequence that differs in zero places from $\vec{k}$ (the constant sequence); in all, there are n such sequences. For each such sequence we obtain $n - 1$ different sequences that differ in one place from it and are still monotonically increasing, and so on. Finally, we take the sum of all sequences that are monotonically increasing and differ in j places from k, which amounts to

$$m \leq \sum_{j=0}^{n-1} \prod_{k=0}^{j} (n-k) = n + n(n-1) + \cdots + n! \leq n(n!)$$

Some sequences, but not more than half, have been counted multiply; thus

$$m \geq \frac{n + n(n-1) + \cdots + n!}{2} \geq \frac{n}{2}(n-1)!$$

Together we have that the number of regular operators $R(n)$ can be roughly estimated by

$$4n((n-1)! - 2) + 1 \leq R(n) \leq 8n(n! - 1) + 1,$$

but the number of n-valued binary functions is 2^{2^n}, so there are few, combinatorially speaking.

5.4 First-order multiple-valued logics

An open question remaining from the list at the end of Chapter 3 is whether there is a class of signs and quantifiers which has more compact rules than Carnielli's. In the meantime we have developed the sets-as-signs tool.

Obviously, we could adopt Carnielli's approach to our system simply by including in the conclusion all distributions that are associated with one of the truth values occurring in the sign of the premise. The tableau rule for $\{0,1\}\,(\forall x)\phi$, for example, would result in a conclusion that comprises all extensions from the rules for $\{0\}\,(\forall x)\phi$ and $\{1\}\,(\forall x)\phi$. But then in the extensions only signs corresponding to singleton sets would occur and it is difficult to see how one can take any advantage from the notion of sets-as-signs. The distributions corresponding to general quantifiers are simply too unstructured to be of any use in a general setting. On the other hand, for practical purposes there does not seem to be a real need for distribution quantifiers. Therefore, we decided to take a pragmatic approach and only consider the usual quantifiers $\forall$ and $\exists$, as defined in Chapter 2. Consequently, we speak of *the* n-valued first-order logic where the particular choice of propositional connectives does not matter.

This leads to somewhat simpler quantifier rules in some cases, if combined with general signs. For example, take the rule for $\{\frac{1}{2}\}\,(\forall x)\phi(x)$ in three-valued first-order logic. A sound and complete rule would be the following:

$$\frac{\{\frac{1}{2}\}\,(\forall x)\phi(x)}{\{\frac{1}{2},1\}\,\phi(t)}$$
$$\{\frac{1}{2}\}\,\phi(c)$$

where t is any term and c is a new Skolem constant.

This rule is justified by the semantics of $(\forall x)\phi(x)$, which asserts the truth value $\{\frac{1}{2}\}$ to $\phi(x)$ iff for all assignments β of x the value of $v_\beta(\phi(x))$ is either $\{1\}$ or $\{\frac{1}{2}\}$ and for at least *one* assignment it is $\{\frac{1}{2}\}$. This is nothing more than an awkward of expressing the fact that $\{\frac{1}{2}\}$ should be the minimum taken over all $v_\beta(\phi(x))$, but it directly justifies the rule given above. Note that the corresponding rule in Carnielli's system has two extensions containing three formulas, not counting the rules that enumerate the used parameters.

It is interesting, however, to note that sets-as-signs can be used to establish a much more compact rule representation of condition 2 given on page 27 than the one suggested by Carnielli (1991).

Recall that Carnielli's solution was to introduce additional rules that list exhaustively the possible truth values of formulas with parameters introduced at earlier stages of the proof. Having the device of sets-as-signs at hand we can formulate condition 2 given on page 27 in a much more natural and compact way by a single signed formula, namely by $\{i_1,\ldots,i_k\}\,\phi(t)$.

Example 5.24. *We continue Example 3.19. The rules for the sign $\{1\}$*

now become:

$$\frac{\{1\}\,(Q\,x)\phi(x)}{\{0\}\,\phi(c_1)} \qquad \frac{\{1\}\,(R\,x)\phi(x)}{\{0\}\,\phi(c_3)}$$
$$\{\tfrac{1}{2}\}\,\phi(c_2) \qquad\qquad \{1\}\,\phi(c_4)$$
$$\{0,\tfrac{1}{2}\}\,\phi(t_1) \qquad\qquad \{0,1\}\,\phi(t_2)$$

where the c_i are new and the t_i are arbitrary parameters.

It is not too difficult to prove completeness of the first-order tableau systems generated by the new rules along the lines given by Carnielli (1991), but we omit it here, since it is rather technical and we do not wish to investigate general distribution quantifiers further in this book.

To conclude this issue, in Figure 5.5 we give a tableau proof of the inconsistency of the set $\Phi = \{1\,(Q\,x)p(x), 1\,(R\,x)p(x)\}$, which can now easily be obtained using the new rules:

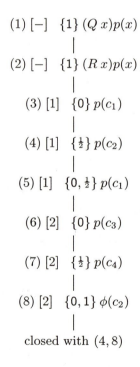

$$(1)\ [-]\ \ \{1\}\,(Q\,x)p(x)$$
$$|$$
$$(2)\ [-]\ \ \{1\}\,(R\,x)p(x)$$
$$|$$
$$(3)\ [1]\ \ \{0\}\,p(c_1)$$
$$|$$
$$(4)\ [1]\ \ \{\tfrac{1}{2}\}\,p(c_2)$$
$$|$$
$$(5)\ [1]\ \ \{0,\tfrac{1}{2}\}\,p(c_1)$$
$$|$$
$$(6)\ [2]\ \ \{0\}\,p(c_3)$$
$$|$$
$$(7)\ [2]\ \ \{\tfrac{1}{2}\}\,p(c_4)$$
$$|$$
$$(8)\ [2]\ \ \{0,1\}\,\phi(c_2)$$
$$|$$

closed with $(4,8)$

FIG. 5.5. Tableau refutation of Φ with complete quantifier rules using sets-as-signs.

Remark 5.25. *The rules above are neither of γ- nor of δ-type, of course,* *although they certainly have a γ-flavour due to the universal term in the conclusion, which means that in general they have to be applied more than once.*

From now on we concentrate on the many-valued generalizations of the quantifiers $\forall, \exists$. Our goal is to extend the uniform notation of classical quantifiers to them. A natural question to ask is what happens in the restricted case of regular logics, that is, when only signs of the shape $\boxed{>i}$ and $\boxed{<i}$ are allowed? The answer, which is surprising at first sight, is given in the next theorem.

Table 5.6 *γ and δ component rules*

γ	$\gamma(t)$	δ	$\delta(t)$
$\boxed{>i}\,(\forall x)\phi(x)$	$\boxed{>i}\,\phi(t)$	$\boxed{<i}\,(\forall x)\phi(x)$	$\boxed{<i}\,\phi(c)$
$\boxed{<i}\,(\exists x)\phi(x)$	$\boxed{<i}\,\phi(t)$	$\boxed{>i}\,(\exists x)\phi(x)$	$\boxed{>i}\,\phi(c)$

Theorem 5.26. (First-order regular logics) *Let $\mathcal{L}$ be any n-valued regular first-order logic. Then the tableau proof system with α- and β- component rules from Definition 5.21 and the γ- and δ- component rules from Table 5.6, together with the classical α-, β-, γ- and δ-rule schemata is sound and complete.*

Proof. The proof rests mainly on the following lemma. Apart from this fact that the proof is exactly as in the classical case; one may proceed along the lines given in Fitting (1990b), with the exception of the changes in the propositional part which have been discussed in Sections 4.2 and 4.3. We do not wish to bore the reader with these trivial modifications, so, rather, we list the lemmata and theorems in Fitting (1990b) which have to be fitted into our framework: Proposition 5.5.2, Theorem 5.7.2, Lemma 5.7.5, Lemma 6.3.2, Theorem 6.3.3, Lemma 6.4.2, and Theorem 6.4.3. A nontrivial modification constitutes Lemma 5.27 below, which replaces Propositions 5.4.1 and 5.4.2 in Fitting (1990b).

Lemma 5.27. *Let $\mathcal{L}$ be any n-valued regular first-order logic. We assume, further, that for all first-order structures occurring in the following statements every element of the universe is representable by some term in the language $\mathbf{L}$:*[*]

1. $\boxed{>i}\,(\forall x)\phi(x)$ *is satisfiable iff* $\boxed{>i}\,\phi(t)$ *is satisfiable for any term t.*

---

[*]It is well known in the classical case that one loses no generality when working with Herbrand structures, which fulfil the representability condition trivially. We could also have done this in the present case, since the concept of a Herbrand universe does not interfere with many-valued notions. The proof of the lemma shows that the representability assumption is only needed for parts 1 and 2 and only for one direction.

2. $\boxed{<i}\,(\exists x)\phi(x)$ *is satisfiable iff* $\boxed{<i}\,\phi(t)$ *is satisfiable for any term t.*
3. $\boxed{<i}\,(\forall x)\phi(x)$ *is satisfiable iff* $\boxed{<i}\,\phi(c)$ *is satisfiable for a new Skolem constant c.*
4. $\boxed{>i}\,(\exists x)\phi(x)$ *is satisfiable iff* $\boxed{>i}\,\phi(c)$ *is satisfiable for a new Skolem constant c.*

Proof. 1. (**Only if:**) Let $\boxed{>i}\,(\forall x)\phi(x)$ be satisfiable. Then in some model **M** for some assignment β

$$v_\beta((\forall x)\phi(x)) \in \left\{ i + \frac{1}{n-1}, \ldots, 1 \right\}.$$

By definition, this holds iff

$$\left(\min_{u \in U} v_{\beta_x^u}(\phi(x)) \right) \in \left\{ i + \frac{1}{n-1}, \ldots, 1 \right\}.$$

By definition of min, this value is unique, say

$$\left(\min_{u \in U} v_{\beta_x^u}(\phi(x)) \right) = i_0$$

and $i + \frac{1}{n-1} \le i_0 \le 1$. Thus, for any $u \in U$,

$$v_{\beta_x^u}(\phi(x)) \in \{i_0, \ldots, 1\} \tag{5.10}$$

A straightforward induction on the depth of formulas, which is done exactly as in the classical case, yields for all ground terms $t \in T_{\mathcal{L}}$:

$$v_{\beta_x^{\beta(t)}}(\phi) = v_\beta(\phi\{x \leftarrow t\}) \tag{5.11}$$

Thus, for any term t, by (5.10) and (5.11)

$$v_\beta(\phi(t)) \in \{i_0, \ldots, 1\} \subseteq \left\{ i + \frac{1}{n-1}, \ldots, 1 \right\}.$$

This is the only step where we need the assumption that the sign is of the form $\boxed{>i}$. If not, $v_\beta(\phi(t))$ might not be a member of the sign set.

(**If:**) Let $\boxed{>i}\,\phi(t)$ be satisfiable for every ground term t. Hence, for all t,

$$v_\beta(\phi(t)) \in \left\{ i + \frac{1}{n-1}, \ldots, 1 \right\}.$$

And by (5.11):

$$v_{\beta_x^{\beta(t)}}(\phi(x)) \quad \in \quad \left\{ i + \frac{1}{n-1}, \dots, 1 \right\} \tag{5.12}$$

By assumption, for each $u \in U$ there is a $t_u \in T$ such that $\beta(t_u) = u$. By this fact and (5.12)

$$
\begin{aligned}
\min_{u \in U} v_{\beta_x^u}(\phi(x)) &= \min_{u \in U} v_{\beta_x^{\beta(t_u)}}(\phi(x)) \\
&= \min_{t \in T} v_{\beta_x^{\beta(t)}}(\phi(x)) \\
&\in \left\{ i + \frac{1}{n-1}, \dots, 1 \right\}
\end{aligned}
$$

2. Similar.
3. **(Only if:)** Let $\boxed{<i}(\forall x)\phi(x)$ be satisfiable. Then in some model $\mathbf{M}$ for some assignment β

$$v_\beta((\forall x)\phi(x)) \in \left\{ 0, \dots, i - \frac{1}{n-1} \right\}.$$

By definition, this holds iff

$$\left(\min_{u \in U} v_{\beta_x^u}(\phi(x)) \right) \in \left\{ 0, \dots, i - \frac{1}{n-1} \right\}.$$

By definition of min, this value is unique, say,

$$\left(\min_{u \in U} v_{\beta_x^u}(\phi(x)) \right) = i_0$$

and $0 \le i_0 \le i - \frac{1}{n-1}$. Choose a $u_0 \in U$ such that $v_{\beta_x^{u_0}}(\phi(x)) = i_0$. Now, just as in the classical case, extend $\mathbf{M}$ to $\mathbf{M}'$ by a constant c that does not occur elsewhere and set $I(c) = u_0$ in $\mathbf{M}'$. By this and (5.11) we have (in $\mathbf{M}'$)

$$
\begin{aligned}
v_\beta(\phi(c)) &= v_{\beta_x^{\beta(c)}}(\phi(x)) \\
&= v_{\beta_x^{u_0}}(\phi(x)) \\
&= i_0 \\
&\in \left\{ 0, \dots, i - \frac{1}{n-1} \right\}
\end{aligned}
$$

and we are finished.
(If:) Let $\boxed{<i}\phi(c)$ be satisfiable for a constant c (this time c needs not be new). By definition,

$$v_\beta(\phi(c)) \in \left\{0, \ldots, i - \frac{1}{n-1}\right\}$$

By (5.11),

$$v_{\beta_x^{\beta(c)}}(\phi(x)) \in \left\{0, \ldots, i - \frac{1}{n-1}\right\}$$

Then, by definition of min,

$$\left(\min_{u \in U} v_{\beta_x^u}(\phi(x))\right) \in \left\{0, \ldots, i - \frac{1}{n-1}\right\}.$$

Here we need the fact that the sign has the form $\boxed{<i}$, since $\min_{u \in U} v_{\beta_x^u}(\phi(x))$ can be any value smaller or equal than $v_{\beta_x^{\beta(c)}}(\phi(x))$.

4. Similar.

The proof becomes a little shorter when one works with Herbrand structures; in particular, the representability condition is trivially satisfied and can therefore be omitted.

We close this section by stating in Table 5.7 the tableau rules for the important case where signs are from the set $\{\{0\}, \ldots, \{1\}\}$. To simplify the notation a little we use the following abbreviations for signs.

Definition 5.28.

$$
\begin{aligned}
\boxed{\leq i} &= \{0, \ldots, i\} = [0, i] \cap N \\
\boxed{\geq i} &= \{i, \ldots, 1\} = [i, 1] \cap N
\end{aligned}
$$

The proof of the analog of Lemma 5.27 could be carried out in a similar fashion as above, using elementary properties of max and min; however, alternatively we can employ the following *contraction* and *splitting rule* in our system:

$$
\frac{\begin{array}{c} \mathsf{S}_1\,\phi \\ \mathsf{S}_2\,\phi \end{array}}{\mathsf{S}_1 \cap \mathsf{S}_2\,\phi}
\qquad\qquad
\frac{\mathsf{S}_1 \cap \mathsf{S}_2\,\phi}{\begin{array}{c} \mathsf{S}_1\,\phi \\ \mathsf{S}_2\,\phi \end{array}}
$$

These rules are obviously sound for every choice of $\mathsf{S}_1, \mathsf{S}_2$ and, when added in a fair manner to an already complete system, completeness is certainly preserved. Now we can also give a *syntactic proof* of the rules in Table 5.7.

Take any tableau proof tree **T** containing applications of the rules from Table 5.7. Consider a situation as in the left tree in Figure 5.6. With the help of the splitting rule we go from the tree on the left to the tree in the middle if we observe that $\{i\} = \boxed{\geq i} \cap \boxed{\leq i}$. By applying the usual rules once to $\boxed{\leq i}\,(\forall x)\phi(x)$, twice to $\boxed{\geq i}\,(\forall x)\phi(x)$, and then the contraction

Table 5.7 *Tableau rules for quantifiers with singleton signs*

$$\frac{\{i\}\ (\forall x)\phi(x)}{\boxed{\geq i}\ \phi(t)}$$
$$\{i\}\ \phi(c)$$

$$\frac{\{i\}\ (\exists x)\phi(x)}{\boxed{\leq i}\ \phi(t)}$$
$$\{i\}\ \phi(c)$$

Where c is a new parameter and t is any term.

rule we obtain a derivation of $\{i\}\ \phi(c)$ as shown in the tree on the right. By substituting every rule application of the kind shown on the left in Figure 5.6 by a sequence of rule applications as shown on the right, we can transform any proof containing rules from Table 5.7 into one which is still valid and contains no rule applications from Table 5.7; and similarly for the existential quantifier.

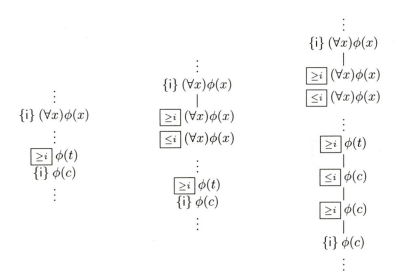

FIG. 5.6. Eliminating universal rule applications with singleton signs.

5.5 Extensions

It is not difficult to see that there are non-regular logics which nevertheless have uniform notation style proof systems. Consider, for example, the implication from Table 2.2 and its rules from Figure 4.3. The reason is that $\supset$ has an asymmetric truth table, whereas regular connectives are always symmetric to a diagonal axis. Would it pay to extend the definition of regular logics to cover such connectives as well? The answer is no, since there is a simpler way to deal with such connectives. They can always be composed from a similar regular connective having appropriate unary regular *filter functions* in one or both of its arguments. Our example $\supset$ can be rewritten as

$$\phi \supset \psi \equiv \nabla\phi \supseteq \psi$$

where

$$\nabla i := \left\{ \begin{array}{ll} 1 & i \neq 0 \\ 0 & i = 0 \end{array} \right.$$

and $i \supseteq j := \max\{i^*, j\}$. Both ∇ and $\supseteq$ are regular. Moreover, substituting $\nabla\phi \supseteq \psi$ for all $\phi \supset \psi$ in a formula results only in linear growth of the formula and proof trees.

Another example of a non-regular connective with uniform style tableau rules is $\bullet$ as defined in Table 5.8.

Table 5.8   *Truth tables of $\bullet$ with two orderings of truth values*

$\bullet$	0	$\frac{1}{2}$	1
0	1	0	1
$\frac{1}{2}$	0	0	0
1	1	0	1

$\bullet$	$\frac{1}{2}$	0	1
$\frac{1}{2}$	0	0	0
0	0	1	1
1	0	1	1

The situation changes when we reorder the truth values in a non-standard way:

$$\frac{1}{2} < 0 < 1$$

The result is the truth table on the right which is regular again with respect to the new ordering.

Let us call a logic **pseudo-regular** if it can be made regular with the help of reordering truth values and use of filter connectives.

Conjecture 5.29. *Every logic with a uniform notation style tableau system is pseudo-regular.*

If the conjecture were true we would have, in a sense, captured *all logics that give rise to a uniform notation style tableau system* with our notion of regular logics. In other words, any logic that is not pseudo-regular cannot have a uniform notation style tableau system then.

However, even if the conjecture is true, there is still room for further extensions. Up to now we have been working with finite totally ordered sets N as truth value sets. Regular signs in this setting were *intervals* in N that included the maximum or minimum of N. It turns out that the notion of a regular sign can be extended naturally to arbitrary partially ordered sets and they coincide with the well-known notions of *upset* and *downset* in the case of lattices.

Definition 5.30. (Partially ordered set) *A* **partially ordered set** *(briefly,* **poset***) is a pair* $\langle P, \preccurlyeq \rangle$*, where P is a non-empty set, $\preccurlyeq$ is a binary relation on P, and (P1)-(P3) below hold:*

(P1) For any $x \in P$: $x \preccurlyeq x$.
(P2) For any $x, y, z \in P$: if $x \preccurlyeq y$ and $y \preccurlyeq z$ then $x \preccurlyeq z$.
(P3) For any $x, y \in P$: if $x \preccurlyeq y$ and $y \preccurlyeq x$ then $x = y$.

Definition 5.31. (Upset, downset) *Let $\langle P, \preccurlyeq \rangle$ be a poset. We define for $a \in P$:*

$$\uparrow a \;=\; \{x \,|\, a \preccurlyeq x, x \in P\}$$
$$\downarrow a \;=\; \{x \,|\, x \preccurlyeq a, x \in P\}$$

For example, if $N = \{0, \frac{1}{n-1}, \dots, 1\}$ and $\preccurlyeq$ is the natural order on N, we have

$$\boxed{\geq i} \;=\; \uparrow i$$
$$\boxed{\leq i} \;=\; \downarrow i$$
$$\{i\} \;=\; \uparrow i \cap \downarrow i$$

and so on. If P is a lattice $\downarrow i$ is a principal ideal and $\uparrow i$ is a principal filter; see, for example, Davey and Priestly (1990).

From this point of view it seems only natural to generalize our framework by admitting a possibly non-linear order on the set of truth values. Besides replacing $<$ and $\leq$ by $\prec$ and $\preccurlyeq$, in Definitions 5.8 and 5.28, one would have to change the definition of the regular connective appropriately and one must also make sure that the technical lemmata still hold under the weaker assumptions. Also one might want to define the conjugation operator in such a way that it soundly reflects a complement within the poset under consideration if negation is to be allowed. In the first-order case one must replace max, min in Definition 2.19(3,4) with appropriate operations on the truth value order.

Natural candidates for a weaker structure to try out would be finite (complemented) lattices or finite (complemented) distributive lattices. These structures have a lot of useful properties which can be used in the technical lemmata.

It might also be interesting to take a look at semi-lattices. All these structures open an easy proof-theoretic path to a kind of logics that is sometimes called *non-linear many-valued logics* (Gabbay, 1991). Note that we can handle such logics in the framework of Chapter 4 already, if we do not insist on uniform notation systems—just treat $\downarrow i$ and $\uparrow i$ as unordered sets of truth values and use them as signs;[*] but, of course, rules that exploit properties of the ordering would be much more concise. Another attractive feature is that the proof system virtually does not change on the surface, although considerable work might be hidden in the proofs of such results as Lemma 5.18 and Lemma 5.20. In Hähnle *et al.* (1993) it is shown that the concept of using signs in connection with ordered structures is not limited to the semantic tableaux framework, but can also be discussed within a resolution framework. It is beyond the scope of this work to explore the connections with ordered algebras in detail here, but we hope to have made it clear that

- (ordered) sets-as-signs, as well as
- regular connectives and signs

are a fruitful and generally applicable concept.

It may be well worth the effort to see how lattice-based logics such as those in da Costa *et al.* (1990), Lu *et al.* (1991), Fitting (1988; 1989; 1990a), and Ginsberg (1986a) can be translated into a tableau framework. Some of the advantages to be gained are the comparability of different approaches within a single framework and the existence of a non-clausal proof system with a readily adapted classical tableau-based theorem prover.

It is worth while noting that propositional regular logics in the infinitely-valued case can be handled using the rules from Section 5.2 without any modification. It is sufficient to observe that in each proof tree only a finite number of signs can occur and that regular signs, by definition, have a finite representation. Recently, a tableau system for fuzzy logic based on the connectives $\vee$, $\wedge$, and $\neg$ of Section 2.3.1, and closed subintervals of $[0, 1]$ as signs was proposed (Kenevan and Neapolitan, 1992). If we add the contraction and splitting rules from above to the rules from Section 5.2 for $\vee$, $\wedge$, and $\neg$, it turns out that Kenevan and Neapolitan's rules (1992) can be easily simulated by them.

In the present chapter we have seen how many-valued first-order logics can be treated nicely in our framework. Other logics, however, are still difficult to handle, even with the extensions described; as an example, take Łukasiewicz logics , which we discuss in Section 6.2.

---

[*]Recently, generalized regular signs, in the sense given above, have been independently rediscovered (Lu *et al.*, 1993) and have been used to characterize the lattice-based logics (called *annotated logics*; cf. also Section 8.1.5 of this book) of Lu *et al.* (1991).

6

BEYOND TABLEAUX

'Of this discourse we more will hear anon.'
— William Shakespeare, *A Midsummernight's Dream*

In this chapter we go beyond the particular framework of semantic tableaux and show that the main concepts developed so far, such as sets-as-signs, regular signs, and connectives can also be used in the context of other inference procedures which are potentially more efficient than are pure tableaux.

The first example we consider is the many-valued variant of a certain refinement of analytic tableaux which is sometimes called tableaux with lemmata or tableaux with restricted cut. It is also closely related to semantic trees and binary decision diagrams (cf. Sections 6.3.2, 8.2.1, and Posegga (1993) and Letz (1993)). We show that this improvement applies to the many-valued case as well as to the classical case.

The other example of a refinement we consider in Section 6.2 also relates to well-known classical concepts. There is a close analogy between classical propositional sentences in CNF and certain 0–1 integer programs. Generating resolvents on the logical side corresponds to generation of cutting planes which prune the solution space of the integer problems.

We present a new technique of how to relate sentences from logic to integer programs which is triggered by tableaux and which not only works for classical propositional sentences in CNF, but also for arbitrary propositional sentences in any finitely-valued logic. Both sets-as-signs and regular signs are crucial in its development.

Finally, in Section 6.3, we briefly mention some other possibilities for exploiting well-known inference techniques from classical logics in a many-valued setting while benefiting from the sets-as-signs technique.

6.1 Lemma generation—asymmetric rules—analytic cut

6.1.1 *Lemma generation and asymmetric rules*

In this section we again take up the discussion thread begun after Proposition 4.5 on page 36.

We have already remarked that in the classical case several logically equivalent tableau rules are possible for most formulas. For example, for the β-rule corresponding to $\mathsf{F}\,(\phi \wedge \psi)$ we have the following possibilities:

$$\frac{\text{F } (\phi \wedge \psi)}{\text{F } \phi \mid \text{F } \psi}$$

$$\frac{\text{F } (\phi \wedge \psi)}{\begin{array}{c|c} \text{F } \phi & \text{F } \psi \\ & \text{T } \phi \end{array}}$$

$$\frac{\text{F } (\phi \wedge \psi)}{\begin{array}{c|c} \text{F } \phi & \text{F } \psi \\ \text{T } \psi & \end{array}}$$

Each rule corresponds to a different covering of the entries equal to 0 in the truth table of conjunction (cf. Example 4.6), as sketched in the diagrams below.

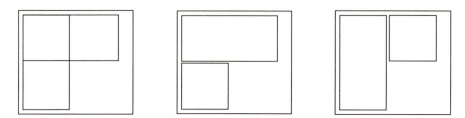

The rule on the left, which is the usual tableau rule, provides, in some sense, less information than the other rules. In each of the asymmetric rules we know more about the truth value of either ϕ or ψ. In theory, this may lead to faster closure of a branch. The problem is, of course, to choose the rule that is most useful. It is difficult to state any general conditions as to when to take either one of the asymmetric rules. In most concrete tableau implementations, however, the problem has to be addressed in any case.

The implementor of a tableau-based theorem prover has to make a strategic decision regarding the search for a closed tableau. Either one expands the tableau until completed and then checks each branch for whether it is closed, or one checks each newly generated branch after every rule application for a closure before putting another formula on an open branch into focus for the next rule application. Both approaches have their pros and cons, but most authors prefer the latter which has the advantage that only one branch (the branch in focus) needs to be retained in the memory at each stage of the proof. We will, therefore, assume an implementation of the latter kind.

Consider the application of the symmetric version of the β-rule to a formula on the current branch. One has to make a decision as to which branch is put into focus for the next rule application. In general, this decision is based on some heuristics. We are not interested in which right now, but it is clear that a decision has to be made somehow. So one of the newly generated branches will be closed first, say the left one. Obviously it would have been better to use an asymmetric rule here, namely the one in the middle which gives us more information about the still open branch.*

*In the first-order case this may cause a problem. Using the asymmetric rules a vast number of additional formulas may be put on the branch which enlarges the search space

This is exactly what we do. Before actually applying a β-rule, decide which branch is put into focus next. Then apply an asymmetric version of the rule which provides more information on the branch last processed.

We call this technique *Lemma Generation*, because if we interpret the additional formula in an asymmetric rule as a lemma the branch that is closed first can be seen as a proof for that lemma. Lemma generation is not merely an *ad hoc* efficiency hack, rather it lifts classical analytic tableaux to a *better proof length complexity class*. To see this, consider the following class of formulas.

Let $p_1, \ldots, p_k$ be propositional variables and let $H_i^k = l_1 \vee \cdots \vee l_k$ denote the clause where l_j is p_j iff the jth digit in the k-ary binary representation of i is 1 and l_j is $\neg p_j$ otherwise. Then define

$$ DA_k = \bigwedge_{i=0}^{2^k-1} H_i^k $$

Informally, DA_k consists of all possible different clauses over $p_1, \ldots, p_k$, where each propositional variable occurs exactly once in each clause.* For $k = 2$ we have

$$ (p_1 \vee p_2) \wedge (p_1 \vee \neg p_2) \wedge (\neg p_1 \vee p_2) \wedge (\neg p_1 \vee \neg p_2) $$

Obviously, all DA_k are unsatisfiable, thus each DA_k has a closed tableau. It was conjectured by D'Agostino (1990) (and later proved in D'Agostino (1992)) that the shortest closed tableau for DA_k has at least $k! \left(1 + \frac{1}{2!} + \frac{1}{3!} + \cdots + \frac{1}{k!}\right)$ interior (non-leaf) nodes, whereas the length of DA_k is merely $\mathcal{O}(2^k)$. Using lemma generation, however, it is easy to show that there are tableaux of polynomial size with respect to the size of the input.

6.1.2   *KE systems and the principle of bivalence*

The class DA_k was originally considered by D'Agostino to support the argument that unmodified tableaux are, in a certain sense, an unnatural proof system, since the DA_k formulas show that the tableau method cannot simulate simple truth table checking in polynomial time, which requires only $\mathcal{O}(k2^{2k})$ steps to check the satisfiability of DA_k.

It is indeed surprising that tableaux can be outperformed by such a simple device as truth table checking; consequently it is not without justification when D'Agostino labels such proof systems as *unnatural*. As a natural alternative to tableaux D'Agostino proposes a system called KE,

---

considerably. Additional strategies have to be applied in order to remove the useless formulas, that is, the formulas that cannot contribute to a proof, as, for example, in Stickel (1992).

*Sometimes, therefore, DA_k is called *the complete formula* of order k.

which was first suggested by M. Mondadori (1988; 1989). Whilst we do not want to go into the details of KE it is illuminating to notice that there is only one rule with more than one extension, namely the *principle of bivalence* (PB), while all other rules are unary. The rules for α-type formulas are as in tableaux and there are two rules for each β-type formula, corresponding to the two asymmetric rules above; see Figure 6.1.

$$
\begin{array}{cccc}
& \dfrac{\beta}{\dfrac{\beta_1{}^*}{\beta_2}} & \dfrac{\beta}{\dfrac{\beta_2{}^*}{\beta_1}} & \dfrac{\alpha}{\begin{array}{c}\alpha_1\\\alpha_2\end{array}} \\
\overline{\mathsf{T}\,\phi \mid \mathsf{F}\,\phi} & & &
\end{array}
$$

where ϕ is any formula and $(\mathsf{T}\,\phi)^* = \mathsf{F}\,\phi, (\mathsf{F}\,\phi)^* = \mathsf{T}\,\phi$.

FIG. 6.1. Principle of bivalence and α- and β-rule schemata of KE.

In KE the DA_k formulas have linear proofs, even if PB is restricted to *analytic* applications, that is, when ϕ is a weak subformula of the theorem to be proved. Between analytic KE and tableaux with lemma generation there is a close connection. Lemma generation can be seen as a kind of restricted principle of bivalence (RPB):[*]

$$
\overline{\beta_i \mid \beta_i{}^*}
$$
where $i \in \{1, 2\}$ and β must already be on the branch.

It is not difficult to see that KE with (RPB) instead of (PB) is p-equivalent to tableaux with lemma generation.[†]

So we have a hierarchy of KE systems with increasingly stronger (PB) rules: KE with (RPB), KE with analytic (PB), and KE with unrestricted (PB). It is an open question, whether the proof length complexity for *arbitrary* propositional formulas of KE with (RPB) (and hence of tableaux with lemma generation) is the same as that of analytic KE. If it were, then KE with (RBP) would have to satisfy the condition ($\mathbf{C}^*$) and ($\mathbf{XF}$) in D'Agostino (1990) and (RPB) seems to be of no help here, precisely because of its restricted occurrence. Hence, we conjecture[‡] that tableaux with lemma generation are weaker than analytic KE, but stronger than tableaux alone.[§]

[*]Note that (RPB) without the proviso is just (PB).

[†]It is crucial for this proof that tableaux are allowed to be closed with non-atomic formulas.

[‡]Recently, Beckert (1992) has suggested a class of formulas which can be viewed as a generalization of DA_k and that seems to separate KE with (RPB) and analytic KE, although there is as yet no proof of this fact. The formula of order n consists of a conjunction of 2^n formulas $(L_1 \wedge q_I) \vee \cdots \vee (L_n \wedge q_I)$, where L_i is either p_i or $\neg p_i$; each conjunct represents one of the 2^n possibilities. I is a string of n 0s and 1s, where the nth digit is 0 if $L_i = \neg p_i$ and 1 if $L_i = p_i$.

[§]For formulas in CNF one easily observes that both notions are (up to polynomial

It is perhaps noteworthy that the power of KE and its relatives depends not so much on the structure of a rule such as (PB) or Cut as on the explicit availability of arbitrary weak subformulas of the theorem to be proved, as the following observation shows. If a signed formula, say ϕ, is to be checked for being a tautology, it suffices to add $(\psi \vee \neg\psi)$ conjunctively for all weak subformulas ψ of ϕ (this does not alter satisfiability and increases the total size of the formula only quadratically) and all resulting theorems θ have shortest proofs in analytic KE that can be p-simulated by the tableau method.

The reason for undertaking this rather lengthy digression on classical proof length complexity is, as D'Agostino (1990, p. 117) rightly observes, that the considerations above carry over to most non-classical logics, and in particular to many-valued logics. In the following we show how lemma generation can be extended to the many-valued case using the sets-as-signs notion. This has similar advantages as it has for the the classical case. Then we sketch a sets-as-signs version of analytic KE.

6.1.3 *Lemma generation in multiple-valued logics*[*]

We are working in propositional logic now for ease of notation and remark that all results also hold in first-order logic.

As is clear from the two-valued case, lemma generation only makes sense when the coverings corresponding to the already processed extensions of a rule and the current extension have a non-empty intersection, since in this case the models excluded by the former extensions can be used to sharpen the current one by putting additional formulas on the branch.

Definition 6.1. (Intersection of extensions) *Let $C_1, \ldots, C_r$ be some extensions of a tableau rule for some formula ϕ. We say that $C_1, \ldots, C_r$* **intersect** *when they have a common model.*

Definition 6.2. (Semantic complement of a set of formulas)   *Let Φ be a set of signed formulas. $\bar{\Phi}$ is a* **semantic complement** *of Φ if the models of $\bar{\Phi}$ are exactly those valuations that are not a model of Φ.*

For lemma generation to make sense we must assume that we are working with strict tableaux now, in particular the newly generated branches after each rule applications are processed in a definite sequence according to a selection rule.

Definition 6.3. (Tableau rule with lemma generation)   *Assume that the tableau rule for ϕ has extensions $C_1, \ldots, C_t$ and that r branches have already been processed using extensions $C'_1, \ldots, C'_r$. If $r = 0$ let C'_1 be*

---

proof length) the same. Therefore, one can argue that in classical logic the distinction between KE with (RPB) and analytic KE is not too important. In many-valued logics, however, where in general we do not have normal forms, it is certainly an issue.

[*]I am much indebted to B. Beckert for working out the details of this section.

the C_{i_1} according to the selection rule. Otherwise, let L be a finite set of direct weak subformulas of ϕ such that

$$\bar{C}'_1 \cup \cdots \cup \bar{C}'_r \cup C_{i_{r+1}} \vDash L$$

where $C_{i_{r+1}}$ is according to the selection rule and define $C'_{r+1} = L \cup C_{i_{r+1}}$.

It is not difficult to show that this definition of a rule results in a sound and complete tableau proof system. For completeness we require the selection rule to be fair.

The choice of L must be such that it excludes intersection between the $\bar{C}'_1, \ldots, \bar{C}'_r$ and $C_{i_{r+1}}$. It turns out that the use of Karnaugh maps is again appropriate here (cf. Section 4.5) to put us on the right track. Consider the Karnaugh map that looks like a truth table for a k-ary many-valued connective (where the leading connective of ϕ, say F, is k-ary), but whose entries are either $-$ or $+$.

Let $\phi = \mathsf{S}\,F(\phi_1, \ldots, \phi_k)$ and S_i be the Cartesian product $\mathsf{S}_{i_1} \times \cdots \times \mathsf{S}_{i_k}$, where

$$\mathsf{S}_{i_j} = \begin{cases} \mathsf{S}' & \text{if } \mathsf{S}'\,\phi_{i_j} \in C'_i \text{ for some } \mathsf{S}' \\ \mathsf{N} & \text{otherwise} \end{cases}$$

Now mark all entries corresponding to a vector in $S_1 \cup \cdots \cup S_r$ with $+$. In a similar way as the S_i, compute an S_{r+1} for $C_{i_{r+1}}$ and mark the entries corresponding to vectors in S_{r+1} with $-$.

The required lemma L is now any extension corresponding to a covering of the Karnaugh map that

1. comprises all negative entries which are not also marked positive, and
2. does not contain at least one entry marked with both $+$ and $-$.

L is not uniquely determined by this procedure, nor is it clear that the required covering can be achieved by *one* extension, but it is easy to see that the first condition assures that Definition 6.3 is satisfied, while the second guarantees that L truly sharpens $C_{i_{r+1}}$. It is a good strategy to choose a minimal covering satisfying the conditions above.

Example 6.4. *Consider the rule for three-valued disjunction and $\{\frac{1}{2}\}$ from Figure 4.3 which we repeat here for convenience:*

$$\frac{\{\frac{1}{2}\}\,(\phi \vee \psi)}{\begin{array}{c|c} \{0, \frac{1}{2}\}\,\phi & \{\frac{1}{2}\}\,\phi \\ \{\frac{1}{2}\}\,\psi & \{0, \frac{1}{2}\}\,\psi \end{array}}$$

Assume that the left extension is processed first and that we need a lemma for the right extension. The Karnaugh map that is yielded by our algorithm is shown in Figure 6.2. The boxed area constitutes the entries needed to be covered (in this case only a single entry), which gives us as a lemma

$\{\{\frac{1}{2}\}\,\phi,\{0\}\,\psi\}$. *A simple subsumption check with the right extension finally gives us the rule*

$$\frac{\{\frac{1}{2}\}\,(\phi\vee\psi)}{\begin{array}{c|c} \{0,\frac{1}{2}\}\,\phi & \{\frac{1}{2}\}\,\phi \\ \{\frac{1}{2}\}\,\psi & \{0\}\,\psi \end{array}}$$

$\vee$	0	$\frac{1}{2}$	1
0		$\boxed{-}$	
$\frac{1}{2}$	$+$	$\pm$	
1			

FIG. 6.2. Karnaugh map for lemma generation.

It is clear that, in a similar way as in Section 4.5, FM techniques can help in finding the required lemmata, although this is perhaps not so straightforward as for the plain rules.

6.1.4   *The principle of multivalence*

We present a *Principle of Multivalence* (PM) and a KE-like system relying on sets-as-signs. To this end we need some more definitions.

Definition 6.5. (Complemented sign and formula) *Let* $S \subseteq N$ *be a sign. The* **complement** *of* S, $\bar{S}$ *is defined as* $N - S$ *which is the set complement in* N. *If* $S\,\phi$ *is a signed formula its complement is defined as* $\bar{S}\,\phi$, *and similar for sets of formulas. Let* $\mathbf{S}_{\mathcal{L}}$ *be the set of signs of some logic* $\mathcal{L}$. *We say that* $\mathbf{S}_{\mathcal{L}}$ *is* **complement closed** *iff* $\bar{S} \in \mathbf{S}_{\mathcal{L}}$ *whenever* $S \in \mathbf{S}_{\mathcal{L}}$.

We use the same notation for the syntactic and semantic complement of a set of formulas. This is justified since the syntactic complement is always a semantic complement.

Definition 6.6. (Many-valued KE) *Let* $\mathcal{L}$ *be any logic with a complement closed set of signs. The* **many-valued KE system** *(MKE) for* $\mathcal{L}$ *is defined by virtue of the following rules.*

Expansion rules. *Let* $\phi \in \mathbf{L}^*$ *be any signed formula and let a tableau rule be defined for it in the sense of Definition 4.4 with extensions* $C_1, \ldots, C_r$. *Then the following are valid rules in the MKE for* $\mathcal{L}$:

$$
\begin{array}{ccc}
\phi & \phi & \phi \\
\bar{C}_1 & \bar{C}_1 & \bar{C}_2 \\
\bar{C}_2 & \vdots & \vdots \\
\vdots & \bar{C}_{r-2} & \bar{C}_{r-1} \\
\bar{C}_{r-1} & \bar{C}_r & \bar{C}_r \\
\hline
C_r & C_{r-1} & C_1
\end{array}
$$

    ···

(Note that if $r = 1$ there is only a single rule, namely $\dfrac{\phi}{C_1}$. Consequently, we obtain the usual α-rules in classical logic when ϕ is of type α. Similarly, if $r = 2$ we obtain the classical KE β-rules.)

Principle of multivalence. *Let $\{S_1, \ldots, S_m\} \subseteq \mathbf{S}_{\mathcal{L}}$ be a covering of N, that is, $S_1 \cup \cdots \cup S_m = N$ and $\phi \in \mathbf{L}$. Then the following is a valid rule in the MKE for $\mathcal{L}$:*

$$
\begin{array}{ccc}
\hline
S_1\ \phi \mid & \cdots & \mid S_m\ \phi
\end{array}
$$

Closure of branches. *Use the contradiction set as defined in Chapter 4.*

The completeness proof for MKE is achieved by combining D'Agostino's proof for KE with the general definitions of the Hintikka set, etc., developed for sets-as-signs. Since it is straightforward we do not elaborate on it.

Example 6.7. *We show the MKE pendant of the proof tree from Example 4.9 in Figure 6.3.*

Remember that α-rule applications are the same as before. Formulas (5) and (7) were generated by an application of (PM). Formula (8) is derived from (7) and (3).

There is, however, a much shorter MKE proof which is shown in Figure 6.4 and which does not even use (PM). This time, (5) is directly inferred from (2) and (3). The reason why we can find this proof is that the application of a MKE extension rule (to formulas (3) and (7) here) can implicitly involve a closure on non-atomic formulas (here $\{0, \frac{1}{2}\}\ \neg p$ and $\{1\}\ \neg p$). If we admit non-atomic closure in many-valued tableaux with lemma generation we can achieve the same effect.

We conclude by remarking that in spite of all that has been said about proof length complexity of KE with (RPB), analytic KE and unrestricted KE can be transferred to MKE. It is not difficult to design many-valued analogues of the separating formula classes.

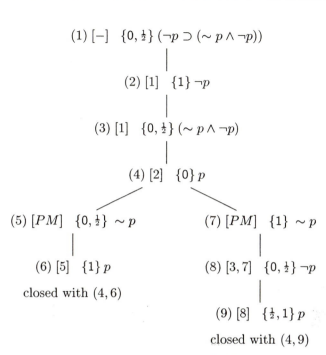

(1) [−] $\{0, \frac{1}{2}\}$ $(\neg p \supset (\sim p \wedge \neg p))$

(2) [1] $\{1\}$ $\neg p$

(3) [1] $\{0, \frac{1}{2}\}$ $(\sim p \wedge \neg p)$

(4) [2] $\{0\}$ p

(5) [PM] $\{0, \frac{1}{2}\}$ $\sim p$ (7) [PM] $\{1\}$ $\sim p$

(6) [5] $\{1\}$ p (8) [3, 7] $\{0, \frac{1}{2}\}$ $\neg p$

closed with $(4, 6)$

(9) [8] $\{\frac{1}{2}, 1\}$ p

closed with $(4, 9)$

FIG. 6.3. A proof in MKE.

6.2   Tableaux as integer programming problems

6.2.1   *Mixed integer programming*

A general MIP problem (Nemhauser and Wolsey, 1989) consists of minimizing a linear function with respect to a set of constraints consisting of linear inequalities in which rational and integer variables can occur. More precise is the following definition.

Definition 6.8. (MIP) *Let $\vec{x} = (x_1, \ldots, x_l)$ and $\vec{y} = (y_1, \ldots, y_m)$ be variables over the rationals and the integers, respectively, and let A, B be integer matrices and h an integer vector. Let $f(\vec{x}, \vec{y})$ be a linear function (called cost function). Then a* **general MIP problem** *is to find $\underline{\vec{x}}, \underline{\vec{y}}$ such that*

$$f(\underline{\vec{x}}, \underline{\vec{y}}) = \min\{f(\vec{x}, \vec{y}) | A\vec{x} + B\vec{y} \geq h\}$$

The general case can be restricted for the present purposes:

- We are interested only in the **feasibility** part of an MIP problem. Nevertheless, we always speak of MIP problems in the following, although we never need to minimize a cost function. This is justified, since, following Hooker (1988) one can always rewrite such problems as minimization problems.

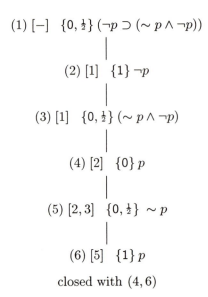

$$(1) \ [-] \quad \{0, \tfrac{1}{2}\} \ (\neg p \supset (\sim p \wedge \neg p))$$

$$(2) \ [1] \quad \{1\} \ \neg p$$

$$(3) \ [1] \quad \{0, \tfrac{1}{2}\} \ (\sim p \wedge \neg p)$$

$$(4) \ [2] \quad \{0\} \ p$$

$$(5) \ [2, 3] \quad \{0, \tfrac{1}{2}\} \ \sim p$$

$$(6) \ [5] \quad \{1\} \ p$$

closed with $(4, 6)$

FIG. 6.4. A MKE proof that illustrates how many-valued KE rules can shorten a proof even without the use of (PM).

- We deal with *bounded* MIP (bMIP) problems; more precisely, all solutions will be in the rational interval $[0, 1]$. If there are no rational variables present we have a bounded integer programming (bIP) problem. In the case where solutions are only from the set $\{0, 1\}$ we speak of a $0 - 1$ IP problem. If only rational variables are present we have a linear programming (LP) problem. It is well known that LP is solvable in deterministic polynomial time.

Lemma 6.9. *bMIP is NP-complete.*

Proof. bMIP is NP-hard, since $0 - 1$ IP already is (Karp, 1972). On the other hand, bMIP $\in$ NP, since it suffices to guess the bounded (by a constant) vector $\vec{y}$ and solve the resulting linear program in NP (in fact in P) time.

Definition 6.10. (bMIP-representable (Jeroslow, 1988))   *A set* $M \subseteq [0, 1]^k$ *is* **bMIP-representable** *if there is a bMIP* (A, B, h) *with* k *rational and* m $0 - 1$ *variables such that*

$$M = \{\vec{x} \mid \text{there is } \vec{y} \in \{0, 1\}^m \text{ such that } A\vec{x} + B\vec{y} \geq h\}$$

It will be convenient to stretch the usual terminology a little. A bIP problem is normally a MIP problem where solutions are integer and from some interval $[-a, b]$. In the following we understand that a bIP problem has solutions in a finite set of rational numbers, more precisely in $\mathbf{n} = \{0, \frac{1}{n-1}, \ldots, 1\}$ for some n. It is trivial to transform such bIP problems into equivalent ones in the usual sense.

Fairly recent overviews on MIP research can be found in Jeroslow (1988), Nemhauser and Wolsey (1989), and Salkin and Mathur (1989).*

6.2.2 Tableau proofs with constraints

In the following we restrict our attention to sets of signs S and many-valued connectives such that the following hold

1. $\mathsf{S} = ([i_1, j_1] \cup \cdots \cup [i_m, j_m]) \cap N$ where $[i_k, j_k]$ are rational intervals (so that each signed formula can be represented by at most $2m$ regular signs).

2. For each such S and k-ary connective F the set $f^{-1}(\mathsf{S})$ must be bMIP-representable.

Note that finitely-valued logics are not restricted by these conditions. Moreover, we assume from now on that all connectives are at most binary.

From now on until further notice let N be finite.

Consider a signed formula $\phi = \boxed{\geq i} F(\phi_1, \phi_2)$, where F is any two-place connective. A signed formula of this type is satisfiable iff for some valuation v the value of $v(F(\phi_1, \phi_2))$ is greater than or equal to i. The key idea in the following is to leave i as well as the signs in the rule extensions uninstantiated. For example, we could write down a rule such as

$$\frac{\boxed{\geq i} F(\phi_1, \phi_2)}{\boxed{\geq i_1} F(\phi_1, \phi_2)}$$
$$\boxed{\geq i_2} F(\phi_1, \phi_2)$$

For most instances of i, i_1, i_2, however, such a rule does not properly reflect the semantics of F, hence we must impose some additional constraints. Let us become a little more precise and consider a signed formula $\boxed{\leq i} (\phi_1 \supset \phi_2)$, where $\supset$ denotes n-valued Łukasiewicz implication (cf. Definition 2.31)

If $i = 1$ the signed formula $\boxed{\leq i} (\phi_1 \supset \phi_2)$ is trivially satisfied and can be omitted from the further analysis of the current branch; otherwise we have the following proposition

Proposition 6.11. If $i < 1$ or, equivalently, $i \leq \frac{n-2}{n-1}$, then $\boxed{\leq i} (\phi_1 \supset \phi_2)$ is satisfiable iff both $\boxed{\geq i_1} \phi_1$ and $\boxed{\leq i_2} \phi_2$ are satisfied by the same valuation and $i = 1 - i_1 + i_2$ holds.

---

*In the first work some connections to automated deduction in classical logics are explained. Recently, some very fast satisfiability checking algorithms for classical propositional logic have been designed in connection with IP, see Hooker (1988; 1991), Jeroslow and Wang (1990) and Hooker and Fedjki (1990). Later we will see that these results might also be useful in the present context.

Proof.   Only If: Assume that $\boxed{\leq i}\,(\phi_1 \supset \phi_2)$ is satisfiable, thus $v(\phi_1 \supset \phi_2) \leq i$. By assumption, $i \leq \frac{n-2}{n-1}$, hence by definition of $\supset$ $v(\phi_1 \supset \phi_2) = 1 - v(\phi_1) + v(\phi_2) \leq i$. Using elementary properties of linear inequalities we have that $1 - v(\phi_1) + v(\phi_2) \leq i$ iff $1 - i_1 + i_2 = i$ and $v(\phi_1) \geq i_1, v(\phi_2) \leq i_2$. Applying the definition of $\boxed{\leq}$ and $\boxed{\geq}$ yields the result.
   If: Similar.

From this proposition we may derive a tableau rule for $\boxed{\leq i}\,(\phi_1 \supset \phi_2)$. It has *provisos* which have to be satisfied in any proof in which the rule is used. A similar technique is used in the classical quantifier rules, but there the proviso can be checked immediately in the case of quantifier rules, whereas we delay the check until tableau completion in the present case. Moreover, different constraints may be associated with each extension. Let us call rules of the new kind **constraint rules**. The rule for $\boxed{\leq i}\,(\phi_1 \supset \phi_2)$ and its counterpart for $\boxed{\geq i}\,(\phi_1 \supset \phi_2)$ are given in Table 6.1. Note that the left extension of the rule for $\boxed{\leq i}\,(\phi_1 \supset \phi_2)$ is empty; only the constraint information $i = 1$ (in which case $\boxed{\leq i}\,(\phi_1 \supset \phi_2)$ is trivially satisfied) counts in the corresponding branch.

Remark 6.12. *It is possible to weaken the constraint to $1 - i_1 + i_2 \leq i$ in the rule for $\boxed{\leq i}$ and to $1 - i_1 + i_2 \geq i$ in the rule for $\boxed{\geq i}$, and the rules still to remain sound.*

Table 6.1   *Constraint rules for $\boxed{\leq i}\,(\phi_1 \supset \phi_2)$ and $\boxed{\geq i}\,(\phi_1 \supset \phi_2)$.*

	$\boxed{\leq i}\,(\phi_1 \supset \phi_2)$			$\boxed{\geq i}\,(\phi_1 \supset \phi_2)$	
	$\boxed{\geq i_1}\,\phi_1$	$i \leq \frac{n-2}{n-1}$		$\boxed{\leq i_1}\,\phi_1$	
$i = 1$	$\boxed{\leq i_2}\,\phi_2$	$1 - i_1 + i_2 = i$		$\boxed{\geq i_2}\,\phi_2$	$1 - i_1 + i_2 = i$

Recall that the number of extensions in n-valued Łukasiewicz implication rules was up to n with the old rules, while now it is constant for arbitrary n. Different values of n are handled in the constraints. Proof trees for the same formula and different N are identical modulo the constraints.
   It is instructive to instantiate the premise of a rule and compute all solutions of its constraint system. Consider, for example, the rule for $\boxed{\geq \frac{1}{2}}\,(\phi_1 \supset \phi_2)$ in five-valued Łukasiewicz logic. The constraint system consists of the single equation $1 - i_1 + i_2 = \frac{1}{2}$, which has to be solved over N. The pairs $(i_1, i_2) \in N^2 = \{0, \frac{1}{4}, \frac{1}{2}, \frac{3}{4}, 1\}^2$ solving this equation are $\{(\frac{1}{2}, 0), (\frac{3}{4}, \frac{1}{4}), (1, \frac{1}{2})\}$.
   Each different solution of the constraint system corresponds to a conventional rule extension, so we can backtranslate our example into a conventional rule with three extensions and no constraints:

$$\boxed{\geq\tfrac{1}{2}}\,(\phi_1 \supset \phi_2)$$

$\boxed{\leq\tfrac{1}{2}}\,\phi_1$	$\boxed{\leq\tfrac{3}{4}}\,\phi_1$	$\boxed{\leq 1}\,\phi_1$
$\boxed{\geq 0}\,\phi_2$	$\boxed{\geq\tfrac{1}{4}}\,\phi_2$	$\boxed{\geq\tfrac{1}{2}}\,\phi_2$

Eliminating the trivially satisfiable formulas $\boxed{\geq 0}\,\phi_2$ and $\boxed{\leq 1}\,\phi_1$ yields exactly the same rule for $\boxed{\geq\tfrac{1}{2}}\,(\phi_1 \supset \phi_2)$ as the sets-as-signs technique from Chapter 4, namely the following:

$$\{\tfrac{1}{2}, \tfrac{3}{4}, 1\}\,(\phi_1 \supset \phi_2)$$

$\{0, \tfrac{1}{4}, \tfrac{1}{2}\}\,\phi_1$	$\{0, \tfrac{1}{4}, \tfrac{1}{2}, \tfrac{3}{4}\}\,\phi_1$
	$\{\tfrac{1}{4}, \tfrac{1}{2}, \tfrac{3}{4}, 1\}\,\phi_2$ $\{\tfrac{1}{2}, \tfrac{3}{4}, 1\}\,\phi_2$

The constraint $1 - i_1 + i_2 = \tfrac{1}{2}$ is merely an implicit representation of the extensions in the conventional rule.

Before we come to the question of branch closure involving signs of the new kind, observe that with each branch of a completed tableau which is not yet tested for closure, a system of linear inequalities whose variables range over N, in other words a bIP problem over N, is associated.

Regarding the detection of closed branches we note, first, that in a completed tableau it is sufficient to look for atomic closure. Also, it is sufficient to look for *pairs* of contradictory formulas (instead of tuples) when signs are regular. In particular, closure can only occur between atomic formulas with different types of sign, that is, it can never occur between formulas such as $\boxed{\leq i_1}\,\phi$ and $\boxed{\leq i_2}\,\phi$. Thus, a branch can only be closed when either two atomic formulas $\boxed{\leq i_1}\,\phi$, $\boxed{\geq i_2}\,\phi$ are present or when there is a single self-contradictory formula $\boxed{\leq j}\,\psi$ (resp. $\boxed{\geq j}\,\psi$).

In the first case, the branch is closed iff the signs have an empty intersection iff $i_1 < i_2$. In the second case, consider $\boxed{\leq j}\,\psi$. There must be a greatest j_0 such that no rule for $\boxed{\leq j_0}\,\psi$ is defined. Then, obviously, no rule for $\boxed{\leq j}\,\psi$ is defined iff $j \leq j_0$ iff $j < j_0 + \tfrac{1}{n-1}$. The value of j_0 is easily obtained from the truth table of the top level connective of ψ. For $\boxed{\geq j}\,\psi$ we proceed similarly.

We thus obtain for each tableau branch $\mathbf{B}_l$ a set of strict linear inequalities

$$\{c_{l_1} i_{l_1} < d_{l_1}, \ldots, c_{l_p} i_{l_p} < d_{l_p}\},$$

such that solving *any* of them results in the closure of $\mathbf{B}_l$; in other words, a disjunction of strict linear inequalities. By definition, a branch is open when it cannot be closed. If we negate the disjunction

$$\bigvee_{k=1}^{p} (c_{l_k} i_{l_k} < d_{l_k}),$$

apply DeMorgan's law, and observe that $c_{l_k} i_{l_k} \not< d_{l_k}$ iff $c_{l_k} i_{l_k} \geq d_{l_k}$ we arrive at the following definition.

Definition 6.13. *Let* **T** *be a completed tableau for* Φ *built up using constraint rules and let* $\mathbf{B}_1, \ldots, \mathbf{B}_m$ *be the branches of* **T**. *Moreover, let* $C_l I_l \leq D_l$ *be the bIP problem (that is,* $\bigwedge_{k=1}^{p} c_{l_k} i_{l_k} \geq d_{l_k}$*) corresponding to the closure of* $\mathbf{B}_l$ *as above. Then* $\mathbf{B}_l$ *is called* **open** *iff* $C_l I_l \leq D_l$ *has a solution that also solves the bIP problem* $A_l I_l \leq B_l$ *associated with the provisos on* $\mathbf{B}_l$. **T** *is a* **constraint tableau proof** *of* Φ *iff it has no open branch.*

Actually, there is another, in some sense simpler, way to represent branch closure. If we view atomic formulas (that is, propositional variables) as object variables ranging over the set of truth values we can take advantage of the fact that the (meta) variables in the signs and (object) variables are of the same type and mix them together in a single constraint. If p is atomic and $\boxed{\geq i}\, p$ is present on the branch $\mathbf{B}_l$ we simply add the constraint $p \geq i$, and similarly for $\boxed{\leq i}\, p$. The resulting constraint system on a branch has then to be solved over $I_l \cup P_l$, where P_l are the propositional variables occurring on $\mathbf{B}_l$. This representation has the advantage of being shorter than the one without object variables, but it has the disadvantage of involving a greater number of variables.*

We summarize.

Theorem 6.14. ϕ *is a tautology iff there is a completed tableau for* $\boxed{\leq \frac{n-2}{n-1}}\, \phi$ *built up using constraint rules which represents a constraint tableau proof.*

6.2.3   *Example*

Before we proceed, let us give an example of a tableau proof using constraint rules. We will give a proof of the formula $p \supset (q \supset p)$ in three-valued Łukasiewicz logic. The conventional proof is shown in the upper part of Figure 6.5, and the proof tree using constraint rules in the lower part. Note that rule application to formula (3) in the lower tree yields only one branch, since adding the condition $i_2 = 1$ makes the constraint system on the branch unsolvable and thus closes it immediately.

The following bIP problem corresponds to the only branch of the tree on the bottom (note that an equation is represented by two inequalities):

---

*This phenomenon is characteristic for MIP representations, see Jeroslow (1988, p. 8).

$$
\begin{array}{llll}
\text{(ia)} & -i_1 +i_2 & & \geq -\tfrac{1}{2} \\
\text{(ib)} & +i_1 -i_2 & & \geq \tfrac{1}{2} \\
\text{(ii)} & -i_2 & & \geq -\tfrac{1}{2} \\
\text{(iiia)} & -i_2 -i_3 +i_4 & & \geq -1 \\
\text{(iiib)} & +i_2 +i_3 -i_4 & & \geq 1 \\
\text{(iv)} & -i_1 & +i_4 & \geq 0
\end{array}
$$

To complete the proof we must show the infeasibility of this problem over $N = \{0, \tfrac{1}{2}, 1\}$.

This may seem not a great achievement as compared to constructing the conventional proof tree, but we must take into account that there exist very efficient algorithms for solving bIP problems and, more important, while the bIP problem *does not become substantially more complex when n grows*, the conventional proof tree becomes bigger and bigger (in the example it grows with $\mathcal{O}(n^2)$, but in general it grows with $\mathcal{O}(n^k)$, where k is the depth of the formula to be proved).

Another important point is that the system of inequalities for each branch can be computed incrementally, while the tree is constructed, thus using information that was computed only once in more than one branch.

6.2.4 *Complexity of multiple-valued logics*

At this point we can ask ourselves which classes of many-valued logics can be axiomatized naturally using constraint tableau rules. Consider the $k+1$-dimensional region R that is defined by each combination of sign variable and k-ary connective f. Let $S(i)$ be one of $\boxed{\leq i}$, $\boxed{\geq i}$. Define

$$
R_f(S(i)) = \{(i, i_1, \ldots, i_k) \mid f(i_1, \ldots, i_k) \in S(i); i, i_1, \ldots, i_k \in N\}
$$

R can be represented as a union or *disjunction* of bIP problems over the $S(i_j)$. Now consider the sum $\Sigma(S(i), f)$ of the maximal numbers of occurrences of each i_j in different problems. This gives us in some sense the amount of copying that is necessary to characterize a connective.

We arrive at a new complexity classification of many-valued logics. Given a logic with many-valued connectives our complexity measure will be a maximum of $\Sigma(S(i), f)$ over all connectives f and signs $S(i)$.

From this point of view the complexity of classical logic, Post logic, and strong Kleene logic is 2, that of classical logic with the equivalence connective is 4, and that of Łukasiewicz logic is again 2. Of course, these numbers will change for different signs than $\boxed{\geq i}$ and $\boxed{\leq i}$. All we can say of the logics mentioned, in terms of traditional complexity classes, is that they are NP-hard; at the moment we do not know how to relate both notions of complexity. We do not want to pursue this topic further here— we are not even sure whether it has an interesting theoretical perspective, but from a practical point of view its mention seems worth while.

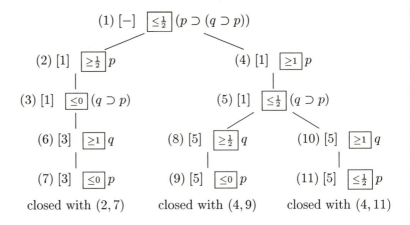

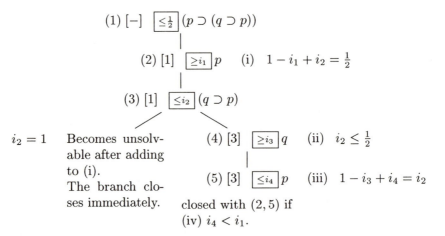

FIG. 6.5. A simple derivation in three-valued Łukasiewicz logic with and without constraints.

6.2.5 *A reduction from multiple-valued deduction to bMIP*

We will now see that constraint tableaux can in fact be linearized, in other words, a tableau can be translated into a single constraint system. In Operations Research, merging of disjunctively connected IP problems has been investigated extensively. Given a set of IP problems over the same set of variables the task is to find a single IP problem, possibly involving auxiliary variables, such that it comprises exactly the disjunction of the solutions of the input systems. The discipline is called **disjunctive programming** and in the general case a constraint system may involve linear inequalities

Table 6.2   *Classical tableau rules in disjunctive
constraint formulation*

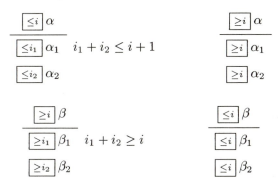

over mixed* variables, hence a bMIP problem.

Our first example will be a formulation of tableau rules for classical propositional logic which consists only of *linear rules* and is illustrated in Table 6.2. Double negations are always eliminated before rule application. Note that only two rules introduce new sign variables and these thus have associated constraints. The bMIP problems resulting from tableaux constructed with these rules are in fact bIP problems, as is the case for all finitely-valued logics.

A prototype implementation of a model checker for classical propositional logic has been implemented using the constraint mechanism of PrologIII, a logic programming language which is able to handle linear inequality constraints. First results are encouraging, for instance the pigeon hole problem for eight holes is solved in about five minutes. The logic capabilities of PrologIII are hardly needed, after the input is transformed into an IP problem; we used PrologIII only for the benefit of having a prototype without having to spend much time on coding. The performance can be increased considerably by experimenting with the IP representation and using a tailor-made IP solving algorithm implemented in C.

Several representations of tableau rules other than the one given in Table 6.2 are possible and some of them pay in increased efficiency of implementation. A variant of the two more complicated of the classical constraint rules where one sign variable is saved is shown in Table 6.3. To derive, for instance, the new β-rule from the old one, simply set i_1 to $i - i_2$ and i_2 to j. The prize is to admit linear expressions as signs, since we have to state the definition of the redundant variable, but this is a mere technical difficulty which does not cause any problems. The rules are complete, even when the constraints in parentheses are omitted altogether. However,

---

*Up until now we had only pure bIP problems where all variables were from a finite set, however, we will soon be in need of the more general case.

Table 6.3 *Improved constraint rules for classical logic*

$$\frac{\boxed{\leq i}\, \alpha}{\boxed{\leq i-j+1}\, \alpha_1 \quad (i \leq j)}$$
$$\boxed{\leq j}\, \alpha_2$$

$$\frac{\boxed{\geq i}\, \beta}{\boxed{\geq i-j}\, \beta_1 \quad (i \geq j)}$$
$$\boxed{\geq j}\, \beta_2$$

Table 6.4 *Improved constraint rules for Łukasiewicz logic*

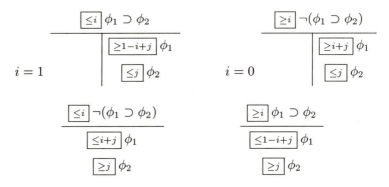

although redundant, the information can reduce the time spent on solving the IP considerably.

A similar optimization is possible for the rules of Łukasiewicz logic. We give them, including the rules for negated implications, in Table 6.4. The rules can be derived from the old rules in a similar way as in the classical case. Again, we can add some constraints which are not necessary for completeness, but could speed up computation. The similarities in the conclusions of the rules suggest that a similar schema as for uniform notation in the classical case may be worked out for Łukasiewicz logics.

6.2.5.1 *Negation* In conventional tableaux there are two possible ways of dealing with negation and the same is true for constraint tableaux and many-valued logics. The first way was chosen in the rules above: negation operators in front of complex connectives are treated together with these connectives as α- or β-formulas, double negations are eliminated, and negated atoms are transformed into inequalities. This works only for some many-valued logics. The second method consists of viewing negations as proper connectives with their own rules, and this works for all kinds of unary connectives. Possible rules for negation (whenever it is defined as $\neg i := 1 - i$) are

$$\frac{\boxed{\le i}\,\neg\phi}{\boxed{\ge 1-i}\,\phi} \qquad \frac{\boxed{\ge i}\,\neg\phi}{\boxed{\le 1-i}\,\phi}$$

6.2.6 *MIP formulation of $\mathcal{L}_\omega$*

In this section we demonstrate that constraint tableaux provide us with a decision procedure even for some infinitely-valued logics. The only approaches to theorem proving in infinitely-valued logics are, to best of our knowledge, these of Beavers (1991) and Mundici (1991; 1993) (cf. also Section 8.2.4). Both are restricted to infinitely-valued Łukasiewicz logic $\mathcal{L}_\omega$. Our approach can handle *any* infinitely-valued logic whose connectives are bMIP-representable; moreover, we claim that it can be implemented much more efficiently. To ease comparison with other approaches we use $\mathcal{L}_\omega$ as an example.

Let us first note that the left-hand rule in Table 6.1 can be reformulated with somewhat weaker constraints, but still sound:

$$\frac{\boxed{\le i}\,(\phi_1 \supset \phi_2)}{\begin{array}{c|c} & \boxed{\ge i_1}\,\phi_1 \\ i=1 & \boxed{\le i_2}\,\phi_2 \quad 1-i_1+i_2 = i \end{array}}$$

The only difference is the removal of $i < \frac{n-2}{n-1}$ in the right-hand side constraint. A moment's thought reveals that this rule is still sound: when $i = 1$ some of the truth table entries in $\boxed{\le 1}$ are now covered in both branches. This is not dramatic, however, since it also happens in the usual classical β-rules.

Next we note that this rule is still sound and complete for $\mathcal{L}_\omega$, since N does not occur in it explicitly.

Applications of disjunctive representation methods (Jeroslow, 1988) and simplification then yields a linear formulation of the same rule:

$$\frac{\boxed{\le i}\,(\phi_1 \supset \phi_2)}{\begin{array}{cc} \boxed{\ge i_1}\,\phi_1 & y \le i, \quad i_1 \le 1-y \\ \boxed{\le i_2}\,\phi_2 & 1-i_1+i_2 = y+i, \quad y \le i_2 \end{array}}$$

where y is binary and i, i_1, i_2 range over N. If $y = 0$ the right extension of the rule above is selected, the left extension if $y = 1$ (since then i_1 is forced to become 0 and i_2 is forced to become 1 which causes the conclusion to become trivially satisfiable). For this reason y is called the **control variable**. The rule on the right in Table 6.1, as well as rules for negation, stays unchanged. To prove the validity of a formula ϕ in $\mathcal{L}_\omega$ we construct a constraint tableau with root $\boxed{\leq c}\,\phi$. Since all rules are linear there is only a single branch from which we extract a MIP problem as before. This problem is then minimized with respect to the linear cost function $f(\ldots, c, \ldots) = c$. If the problem is feasible and the resulting minimum is 1, then we know that $v(\phi) = 1$ must hold in all models v of ϕ, because the constraint tableau is infeasible whenever $v(\phi) \leq c < 1$ is stipulated.

In this way we can translate every many-valued deduction problem from $\mathcal{L}_\omega$ into a single MIP problem whose integer part has no more variables than the input formula has connectives.

Corollary. SAT$_{\mathcal{L}_\omega} \in$ NP.

This result was obtained in Mundici (1987a) in a rather more complicated way using McNaughton's Theorem (McNaughton, 1951). Our completely different method is not only much simpler, but also lends itself to many other logics for which such results as McNaughton's Theorem do not exist.[*]

It is not trivial to compute MIP representations of many-valued tableau rules, as can be seen in the example above, and it also is not easy to solve MIP problems, but the work done in the field of Operations Research (see, for instance Balas *et al.* (1993)), where a considerable amount of knowledge about MIP methods has been accumulated, fits in here exactly.

Successful heuristics for selecting the control variables to be fixed next in branch-and-bound methods (Jeroslow and Wang, 1990; Hooker and Fedjki, 1990; Plaisted and Lee, 1990) can be adapted from the two-valued Davis-Putnam-like procedures in which they are used to the many-valued MIP setting.

As before in the classical case, the rules for Łukasiewicz logic can be simplified:

$$
\frac{\boxed{\leq i}\,\phi_1 \supset \phi_2}{\boxed{\geq 1-i+j-y}\,\phi_1 \qquad y \leq i}{\boxed{\leq j+y}\,\phi_2 \qquad j \leq i}
\qquad\qquad
\frac{\boxed{\geq i}\,\phi_1 \supset \phi_2}{\boxed{\leq 1-i+j}\,\phi_1}{\boxed{\geq j}\,\phi_2}
$$

We close this section with an example. We prove that the formula from the example in Section 6.2.2 is a tautology even in $\mathcal{L}_\omega$. The tableau is

---

[*]The other direction, NP-hardness, was also shown in Mundici (1987a). The idea is to define for each set of propositional variables $p_1, \ldots p_k$ a $\mathcal{L}_\omega$-formula two$(p_1, \ldots p_k)$ such that two$(p_1, \ldots p_k)$ is satisfiable in $\mathcal{L}_\omega$ iff $p_1, \ldots p_k$ are assigned binary truth values. Then for any formula ϕ which contains the propositional variables $p_1, \ldots p_k$ it is true that ϕ is satisfiable in classical logic iff two$(p_1, \ldots p_k) \supset \phi$ is satisfiable in $\mathcal{L}_\omega$. If two has polynomial size in k this property reduces SAT to SAT$_{\mathcal{L}_\omega}$. In $\mathcal{L}_\omega$ the function two is a little awkward to define, since there is no connective corresponding to a truth value set complement. The technique works for many other non-standard logics.

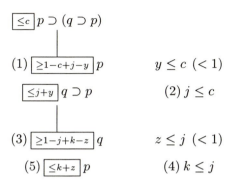

$$\boxed{\leq c}\, p \supset (q \supset p)$$

(1) $\boxed{\geq 1-c+j-y}\, p$ $y \leq c\ (<1)$

$\boxed{\leq j+y}\, q \supset p$ (2) $j \leq c$

(3) $\boxed{\geq 1-j+k-z}\, q$ $z \leq j\ (<1)$

(5) $\boxed{\leq k+z}\, p$ (4) $k \leq j$

Instead of minimizing c we directly show that the problem becomes infeasible if the strict inequality $c < 1$ is added to the system; using $y \leq c$ and $c < 1$ we can fix $y = 0$, since y is a binary variable, which simplifies the resulting MIP. A similar argument goes for z. The numbered inequalities become part of the characteristic MIP for the problem

$$\{c < 1\} \cup \left\{ \begin{array}{llllll}
(1) & & p & -j & & +c & \geq 1 \\
(2) & & & -j & & +c & \geq 0 \\
(3) & q & & +j & -k & & \geq 1 \\
(4) & & & j & -k & & \geq 0 \\
(5) & & -p & & +k & & \geq 0
\end{array} \right\}$$

which is in fact only a LP problem and is easily seen to be infeasible.* Such simplifications due to obvious variable fixings occur quite often and count towards the strengths of the MIP approach.

6.2.7 *Outlook*

We summarize the directions for further research which seem to be the most promising:

- Apply fast satisfiability checkers for many-valued logics to fuzzy reasoning and verification of integrated circuits; see also Section 7.3.2.
- The effect of tableau speed-up methods such as additional inference rules, lemma generation, indexing, etc., on the MIP translation should be investigated. One example: while going from search trees to MIP representations structural information is lost. On the other hand, we might propagate such information from the tableau to the MIP representation, for instance by specifying a partial order on variables which is then used to determine the order in which they are fixed when solving the MIP.

---

*Infer $-p + j \geq 0$ by adding (4) and (5); from this and (1) $c \geq 1$, which contradicts $c < 1$.

- IP is difficult, while LP is not. It has been shown in other contexts that with certain inference rules (for example unit resolution) it is safe to substitute LP for IP. It would be interesting to identify such situations in the many-valued context.

- Perhaps the most interesting and challenging task is the extension to first-order logic. Several approaches are possible and are currently being investigated. First results are reported in Ries and Hähnle (1993).

- The technique may well be applicable to other non-classical logics, such as modal or temporal logics.

A much extended treatment of the material in this section is to appear in Hähnle (1993a).

6.3   Other inference systems

6.3.1   *Techniques common to tableau-like calculi*

In this section we briefly state some techniques for improving classical deductions which are more or less state-of-the-art in tableau-like calculi (the connection method (Bibel, 1987), connection tableaux (Letz, 1993), model elimination (Loveland, 1969), etc.) and which can also be applied in a many-valued setting. Since these techniques are not specific for many-valued deduction we give only very short descriptions and pointers to the literature. The point we want to make is that all these techniques are fully compatible with the techniques for efficient proof search in many-valued logics described in this book.

6.3.1.1   *Proof representation*   Grundy (1990) suggested an efficient data structure for the representation of tableaux which is always polynomial in size of input in the propositional case and thus avoids the space explosion of naïve tableaux. If preprocessing and transformation steps approach a certain complexity we can speak of a *compilation* of the input. In particular, one could try to transform tableaux into definite Horn clauses which are then executed by Prolog. A similar technique has been used successfully for BDDs (cf. Section 8.2.1) by Posegga (1992a). The latter has been extended to full first-order logic (Posegga, 1993). A prototypic implementation is available.

In the theorem prover SETHEO (Schuhmann *et al.*, 1991; Letz *et al.*, 1991), a tableau-like deduction system triggers compilation of first-order knowledge bases in CNF into an Extended Warren Abstract Machine code.

6.3.1.2   *Free variables*   In Smullyan's popular formulation of first-order tableaux which we also adopted in the present framework the γ-rule re-

quires to substitute an arbitrary but fixed term* for the quantified variable, see Table 3.1. Since this 'guess' may be wrong, the γ-rule may have
to be applied again and again on the same universal type formula in a
tableau proof. Obviously, this indeterminism can make proofs very long
and it is appropriate to postpone the instantiation needed in a γ-rule until
more information on the actual instance required has been collected. We
know of three (independent) approaches in the literature where this has
been expressed formally (Reeves, 1987; Fitting, 1990b; Ophelders, 1992).
In recent papers by Hähnle and Schmitt (1993) and Beckert, Hähnle, and
Schmitt (1993) these results have been improved considerably. The results
carry over to the many-valued case without any restrictions or modifications. More generally, the techniques for proof search in many-valued logics
described in this book are fully compatible with unification as a technique
for efficient proof search. This fact is important since *any* first-order proof
procedure which claims efficiency involves unification. Consequently, our
ideas carry over, for example, to the connection method or even resolution.
Whereas the first is more or less the same as tableaux, in order to make
use of the notions developed here for the latter some work must be done.
See Section 6.3.4 for some hints.

6.3.2 *Decision diagrams*

We take a closer look at many-valued decision diagrams (see Section 8.2.1
for the basic definitions). Since edges in Shannon graphs have a similar
task as signs in tableaux, it is appropriate to label edges with *sets of truth
values* and achieve a similar effect as with tableaux.

Consider, for example, the rule under Step 1 on page 142. Using sets
of truth values the conclusion becomes

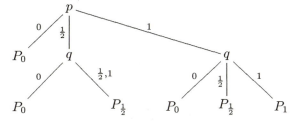

Simplification rules have to be modified accordingly. This is only a
minor improvement. The reason is that unlike the tableau rules, BDD
rules are generated by a column-wise analysis of the truth tables, thus
leaving less room for optimization. We can, however, achieve a much more
concise BDD representation of many-valued logics when we admit variables
on the edges. The rule under Step 1 on page 142 might then become

---

*Smullyan did not include function symbols in his first-order language, so in his case
constants were the only ground terms.

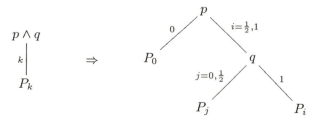

where an edge labelled $i = i_1, \ldots, i_m$ is understood as an abbreviation for m edges labelled $i_1, \ldots, i_m$ and with respective successor nodes $P_{i_1}, \ldots, P_{i_m}$. If these abbreviations are expanded the usual rule results. Using this trick, similar savings as for tableau rules may be achieved. Additional savings are possible when some of the P_is in the premise are identical and this fact is taken into account by the rule.

Note that in the classical case the technique is not very useful, since it is only a kind of shorthand notation for BDDs; thus only the representation size is decreased, but not the number of computation steps. Representation size, however, is usually not a problem in the classical case, while it can be a considerable one in the many-valued case.

In Bryant (1986; 1992) and Posegga (1992a) it was demonstrated that when using sophisticated programming techniques classical BDDs are quite a powerful decision procedure for classical propositional logic. Posegga (1993) showed that the framework can be lifted to first-order logic.

We suspect that some, though not all,* of the work done in the BDD area can be paralleled for many-valued logics.

6.3.3 *Extending multiple-valued dissolution*

The many-valued dissolution rule of Murray and Rosenthal (see Section 8.2.3 for the basic definitions) is limited to a relatively narrow class of many-valued logics, so-called UNF logics. In the following we sketch a technique that enables the application of dissolution to arbitrary finitely-valued logics.

The key idea is to use our sets-as-signs tableau rules to transform a many-valued logic formula into a UNF formula. From then on the method discussed in Section 8.2.3 can be applied.

We saw earlier (in Section 4.5) that the tableau rules employing sets as signs are nothing other than a minimal SOP representation of the characteristic function associated with a sign and connective. Thus, given a complex signed formula S ϕ, its tableau rule determines a representation of S ϕ where only signed direct subformulas of ϕ and conjunctions and disjunctions occur.

---

*Some results for BDDs rely upon normal form properties (Bryant, 1992) which may not be present in many-valued logics.

As an example take the signed formula $\{\frac{1}{2}, 1\}$ $(\phi \cong \psi)$ from $\mathcal{L}^3_{M+}$ (cf. Definition 2.29, Figures 4.3 and 4.4). The tableau rule gives us

$$\{\tfrac{1}{2}, 1\} \, (\phi \cong \psi) \equiv (\{0\} \, \phi \wedge \{0\} \, \psi) \vee (\{\tfrac{1}{2}\} \, \phi \wedge \{\tfrac{1}{2}\} \, \psi) \vee (\{1\} \, \phi \wedge \{1\} \, \psi)$$

Hence, by recursive descent through a signed formula we may obtain an equivalent formula which is built up completely from conjunctions, disjunctions, and signed atoms.* This formula has the form of a UNF formula after the preprocessing step where unary connectives have been substituted by signs.

The technique presented restores dissolution as a general proof procedure for many-valued logics.†

Moreover, in a similar fashion other path-oriented proof procedures such as the connection method of Bibel (1980; 1987) or the matings of Andrews (1981; 1986) can be generalized to cover many-valued logics.

The above also applies to other results of previous chapters, such as the theorems on regular logics and first-order many-valued logics.

On the other hand, the techniques developed for MIP representation of analytic tableaux do not easily carry over to any of the other proof procedures mentioned. The reason is that the tableau procedure has a clear rule-based description and at the same time the tableaux have a simple logical structure. The other proof procedures lack one of these properties.

6.3.4 *Resolution*

In this section we take the point of view that an efficient resolution system requires a kind of CNF. In Section 8.1.3 we review an approach to many-valued theorem proving which is based on non-clausal resolution, but it remains to be seen whether this is viable. In Sections 8.1.1 and 8.1.2 some specialized attempts for achieving CNF are reported. A more general approach is taken in Baaz and Fermüller (1992), but the CNF algorithms in all these papers have in common the fact that they produce clause sets which are much more redundant than necessary.

There are three main obstacles to be overcome when clausal normal forms are to be used in a generalized context:

1. Normal forms can become exponentially long wrt the length of the input when a naïve algorithm is used. This is not so problematic in classical logic where knowledge bases usually consist of conjunctions of relatively short formulas. In non-classical logics, however, even relatively short formulas can become quite long during this process. In Hähnle (1992b, pp 17–19) an example that illustrates the situation in many-valued logics can be found.

---

*It is obvious how satisfiability for such formulas has to be defined.

†A similar technique has been suggested independently by Murray and Rosenthal (1991c; 1993b).

2. The normalized input bears no resemblance to the original formula. This makes it difficult to explain the machine-generated proof to the user.

3. Many non-classical logics may fail to have normal forms, or at least it is non-trivial to find them.

The first two problems can largely be solved by using a *structure preserving clause form translation* (Tseitin, 1970; Boy de la Tour, 1990; Eder, 1991), which has the double advantage of

1. producing normal forms in linear time and space wrt to the input, and

2. establishing a relationship between the clauses of the normal form and the subformulas of the input formula.

As to the third problem, it is not likely that there is a uniform solution to it due to the diversity of non-classical logics. It has been shown, however, that for certain classes of non-classical logics structure preserving CNF transformations (and corresponding resolution rules) can be devised to give the desired results (Mints, 1990). These include intuitionistic logic and the modal logics S5, S4, T, K, K4.

The central idea behind structure preserving clause form translations is to introduce additional atoms which serve as abbreviations for subformulas of the input. Assume that we are working in classical logic, that we have a propositional formula ϕ, and that we need a finite set of clauses X_ϕ such that $\vDash \phi$ iff $X_\phi \vdash \square$.

Let $SF(\phi)$ denote the set of subformulas of ϕ (note that the size of $SF(\phi)$ is linear wrt the length of ϕ) and let $n = |SF(\phi)|$. We denote by L^* the complement of a literal, cf. Definition 5.1. Now we introduce a new variable p_i for each $\phi_i \in SF(\phi)$ which is not a literal and consider for each $1 \leq i \leq n$ and $\phi_i = (\phi_j \text{ op } \phi_k)$ the formula

$$p_i \cong (p_j \text{ op } p_k) \tag{6.1}$$

where op is the top-most connective of ϕ_i and p_j, p_k either correspond to ϕ_j, ϕ_k or $\phi_l = p_l$ if ϕ_l is a literal. This process is called *abbreviation*, *definition*, or *renaming* by various authors. Let $X_{\phi,i}$ be a CNF representation of (6.1). Now we can define

$$X_\phi = \left(\bigcup_i X_{\phi,i} \right) \cup \{p_1^*\}$$

where p_1 is the definition of ϕ. It is fairly easy to see that X_ϕ indeed has the desired properties.

Additional savings are possible by analysing the polarity of the subformulas and substituting either $\subset$ or $\supset$ for $\cong$.

In the many-valued case we work with **signed clauses**, i.e., the (classical) clauses over $\mathbf{L}_0^*$.

In order to turn a signed formula into a set of signed clauses we use *inverse* tableau rules (Kapetanović and Krapež, 1989), where extensions are conjunctively connected and the extensions themselves are signed clauses.

It is possible to define a sound and complete resolution rule on signed clauses, for instance,

$$\frac{\mathsf{S}_1\, p \vee D_1 \quad \cdots \quad \mathsf{S}_m\, p \vee D_m}{D_1 \vee \cdots \vee D_m} \qquad \text{if } \mathsf{S}_1 \cap \cdots \cap \mathsf{S}_m = \emptyset$$

together with an appropriate factoring rule.

In Hähnle (1993b) the ideas which are only sketched here are elaborated and a many-valued generalization of polarity is defined which helps to remove redundant signed clauses.

6.4   Evaluation

6.4.1   *Evaluating the criteria from the introduction*

In the Introduction we stated a catalogue of desirable properties of a framework for many-valued theorem proving. Let us go through this catalogue and check to what degree the framework proposed in this book fulfils those properties.

- **Wide applicability** For propositional logics we have certainly achieved this particular goal; for quantified logics we could provide an efficient treatment for the many-valued generalizations of $\forall$ and $\exists$. We showed that even for other quantifiers the sets-as-signs technique offers a considerable improvement over previous approaches. On the other hand, if quantifiers other than $\forall$ and $\exists$ really were needed in an application, more work would certainly have to be done. The results of Section 6.2 show that it is possible to deal with infinitely-valued propositional logics.

- **Flexibility** Due to the generic soundness and completeness proofs of Chapters 4 and 5 we have a truly flexible framework.

- **Easy adaptability** Based on the ideas presented in this book a tableau-based theorem prover for finitely-valued logics has been built (Hähnle *et al.*, 1992) which does possess exactly the kind of modularity required for easy adaptability to different logics (see also the next section).

- **Performance** Regarding propositional logics, excellent performance can be expected from an MIP-based implementation (cf. Section 6.2). Using the quantifier rules of Chapter 5 such an implementation may be lifted to first-order logic by a similar method as that considered by Lee and Plaisted (1992). For the time being, the theorem prover

$_3TAP$ (Hähnle *et al.*, 1992) may be used with acceptable performance. Also, good performance can be expected from an implementation of the resolution principle on signed formulas (see Section 6.3.4 and Hähnle (1993b)).

- **Closeness to classical versions** We hope that in the preceding sections we have made it clear that most known inference systems and strategies can be extended to many-valued logic using our techniques. The results of Chapter 5 on uniform notation systems in particular show that many-valued reasoning can be done by a relatively small and natural extension of classical techniques.

All in all, the goals we set out have been achieved to a fair extent. Of course, much work can still be done, for example regarding non-standard quantifiers, a resolution implementation, or lattice-based logics. This book is a mere foundation for many-valued theorem proving, but we are now in a position where applications of many-valued logic can actually be tried. And sooner or later the demands of real applications will spin off further refinements on the theoretical side.

In the following section we substantiate our claim that the techniques developed in this book make many-valued theorem proving a real possibility by reporting some experiences with an experimental implementation.

6.4.2 *Some experimental results*

From 1990 to 1992 at the University of Karlsruhe the experimental tableau-based many-valued theorem prover $_3TAP$ was developed as part of the TCG joint project between the University of Karlsruhe and the IBM Germany Science Center in Heidelberg.[*]

A detailed description of $_3TAP$ may be found in Hähnle *et al.* (1992), for a short survey (written in German) see Hähnle (1990a).

$_3TAP$ can in principle handle arbitrary finitely-valued first-order logics and makes use of the techniques and results developed in Chapters 4 and 5. To make it capable of handling a new logic, all one has to do is to change the rule base and perhaps add some declarations of signs and connectives. Among others, the logics mentioned in Sections 7.3.1 and 7.3.2 were implemented.

In order to gain some insight into how much is achieved in practice using the sets-as-signs concept, two implementations of the three-valued logic used in Gerberding (1991) were provided. The first version runs with rules derived from Carnielli (1987), that is, only singleton signs occur. The second version has a full set of signs and rules, that is, for each $\emptyset \subsetneq S \subsetneq \{0, \frac{1}{2}, 1\}$ and connective a rule is computed (for some combinations no rule

---

[*] $_3TAP$ is available without charge to research institutions. Please contact the author if you are interested in receiving a copy. $_3TAP$ is written mainly in Prolog.

Table 6.5 *Some Test Results using ${}_3\mathcal{T\!A\!P}$*

Problem	Closed branches when using			
	Carnielli's method		Sets-as-signs	
	Unlinked*	Linked†	Unlinked	Linked
Lemma 5.1	225	119	7	7
Theorem 5.3	225	82	23	13
Lemma 5.9	—‡	33	4	4
Figure 5.14	254	254	10	8
Figure 5.17	—	—	—	8
Axiom MVEQ4	46	33	6	4

is defined, however). In Table 6.5 statistical figures of runs in both versions are summarized for various first-order problems from Gerberding (1991).

Run times are roughly proportional to the number of branches generated, thus there is a very clear advantage in the sets-as-signs approach. All run times required in this case are within fractions of a second. Note that the sample problems are not particularly difficult and the underlying logic has only three truth values. For logics with a greater number of truth values the difference becomes even more spectacular.

6.4.3 *Outlook*

Considering the material developed in the previous chapters and the discussion in the present section, what recommendations concerning the logical basis can we give to someone who is willing to build a many-valued theorem prover for real applications?

The answer mainly depends on the expressive power of the logics under consideration and, to a certain degree, on the application. For propositional infinitely-valued logics an MIP-based implementation as sketched in Section 6.2 is probably best. For other logics, the state-of-the-art approach in the two-valued case should be taken and modified using the sets-as-signs technique. In particular, for propositional finitely-valued logics a modification of the Davis-Putnam procedure seems best if tautology checking is the aim (Buro and Büning, 1992) and BDDs if simplification of large and unstructured expressions is desired (Bryant, 1992). For first-order finitely-valued logics a resolution framework as sketched in Section 6.3.4 (see also Hähnle (1993b)) is very interesting. In the future refinements of tableaux might also turn out to be competitive (Letz *et al.*, 1991; Letz, 1993).

---

*All formulas are present on the initial branch.

†Only formulas with links into the formula in focus are fetched from the knowledge base on demand.

‡No proof found after several minutes.

7

APPLICATIONS

Das Erforschliche in Worte sieben;
Das Unerforschliche ruhig veralbern:
— Arno Schmidt, *Seelandschaft mit Pocahontas*

In this chapter we give an answer to the question

DOES MANY-VALUED THEOREM PROVING HAVE ANY USE?[*]

We begin with a fairly exhaustive list of (possible) applications of many-valued logics. Some of them will be examined more closely in the following two sections. Some of these applications demand a many-valued theorem prover of considerable flexibility.

7.1 Overview

1. Program verification, specification and synthesis
   There are various approaches to reasoning about programs that use three- and four-valued logics to model semantics of partially defined and/or distributed programs.[†] In some way or another statements about programs must be reduced to first-order formulas.[‡] At this point a many-valued theorem prover may be useful. The most important approaches using many-valued logic include

   (a) several groups of researchers concerned with the Vienna Definition Method (VDM), notably the **MetaSoft** project. See Blikle (1991) for an overview and some recent developments.

   (b) Many-valued interpreters are used to assign a proper meaning to negation in logic programming, see, for example Fitting (1985; 1986; 1990a), Fitting and Ben-Jacob (1990), Kunen (1987), and Sheperdson (1989).

2. Artificial intelligence

   (a) In de Bessonet (1991) many-valued logic is used within an AI reasoning system to express various grades of presence of properties.

---

[*]The question is a slight variation of the title of a paper by Scott (1976).

[†]In fact it was the semantics of partial recursive functions which provided the motivation for one of the pioneer papers in the field of many-valued logics (Kleene, 1938).

[‡]In the case of Prolog, programs *are* formulas.

(b) In Ginsberg (1986a; 1986b) many-valued logics with a structured set of truth values are used to express hierarchies and
dependencies in truth maintenance systems.

(c) Belnap's four-valued logic (see Section 7.2.2 below) has been
used as a theoretical basis for knowledge representation languages in Patel-Schneider (1990; 1989) and Bittencourt (1989).

(d) In Schöpke (1990; 1991) an application of many-valued logic for
decision support is described. Truth values are used to model
states in certain situations. Situations are modeled by propositional formulas in many-valued logic. A theorem prover may
help to detect the state of a given situation.

3. Logic

(a) Belnap (1977) used a four-valued logic to model relevance logic.
Recently, D'Agostino gave a tableau system for this logic which
turns out to be a special case of our work (see Section 7.2.2).

(b) Caferra and Zabel (1990) used many-valued logics to implement
a theorem prover for the modal logic S5. This can be improved
and generalized using our results; see Section 7.2.4.

(c) In MacNish and Fallside (1990) a three-valued logic is used to
model default reasoning. The tableau-based default reasoning
approach of Schwind (1990; 1991) can perhaps be made more
fine-grained using many-valued logics. Ginsberg (1986a; 1986b)
shows how to use lattice-based many-valued logics to implement
a kind of default reasoning system.

(d) Doherty (1990a; 1990b; 1991) and Doherty and Lukaszewicz
(1991) used a three-valued logic to model a non-monotonic logic.
In fact the tableau system given in Doherty (1991) is a special
case of the systems from Chapter 4. Recently, very interesting
connections between many-valued logic and non-monotonic reasoning were found by Stachniak (1993).

(e) Many-valued matrices* have long been used in proof theory to
establish the independence of axioms in Hilbert-type inference
systems. Since these proofs are extremely tedious to do by hand
it may be valuable to have a proof assistant based on a flexible
many-valued theorem prover that can quickly check whether a
given matrix does the job. See Section 7.2.1.

4. Natural language processing
The use of many-valued logic to model partial knowledge was sugge-
sted in a number of papers in the area of natural language processing
(NLP), see Blau (1978), Fenstad *et al.* (1985; 1988), and Langholm

---

*The technique is due to Paul Bernays, see Gottwald (1989) and Urquhart (1986) for
an introduction.

(1989). In particular, for dealing with presuppositions (implicit assumptions) some proposals were made to use three-valued logic; see, for example Seuren (1985). In Schöpke (1991) Seuren's logic was axiomatized and tested by a tableau-based theorem prover using our sets-as-signs rules. The gap between theory and application on the linguistic side has yet to be bridged however.

5. Error-correcting codes
   Mundici (1989; 1990) suggested using many-valued logic in the area of adaptive error-correcting codes. He showed that Łukasiewicz's truth values have a natural interpretation, namely as a measure for the degree of refutation of a certain meaning that a piece of information has, provided the number of possible errors does not exceed $n-2$ in an n-valued logic. More complex applications potentially arise in any AI deduction system where one has to deal with a finite, though bounded, number of possible errors in the data.

6. Interval arithmetic
   In the arithmetic of intervals of natural numbers, three-valued models arise naturally, see Section 7.3.1. A many-valued theorem prover can be used to prove properties of such many-valued axiomatizations.

7. Hardware verification
   In hardware design and analysis there is a strong tradition of using many-valued logics for various purposes (ISMVL-21, 1991; ISMVL-22, 1992). Most applications, however, have been either simulation tools or implementations of genuinely many-valued circuits. In Section 7.3.2 we describe briefly how many-valued logics render themselves naturally for the verification of switch-level specifications.

8. Quantum physics
   In Mundici (1986) it is shown that the C^*-algebras of quantum lattice systems can be coded as sets of sentences in the infinitely-valued Łukasiewicz logic. Thus, theorem proving procedures in many-valued logic can be used to make computations over the corresponding C^*-algebras. See, for instance, Mundici (1987b).

7.2   Applications of a theoretical nature

7.2.1   *Independence proofs in Hilbert systems*

Assume that we have a set of axioms and rules which are sound and complete for some strongly finite propositional logic. We suspect that one of the axioms is independent of the others, since we fail to prove it from the others. A technique of proving that it is in fact independent based on many-valued logics goes as follows:

With each connective we associate an appropriate many-valued truth table and define a subset of the truth values as designated. Then we show that all axioms except the one that is to be proved independent take on a designated truth value under all valuations. Next we show that the rules are closed under propagation of designated truth values, that is, whenever the premise has a designated truth value so has the conclusion. The final step is to show that the axiom to be proved independent may take on a non-designated truth value. The proof that this setup guarantees independence may be found, for example, in Gottwald (1989).

The number of truth values of the logic used in the independence proof must be larger than the number of truth values of the object logic. Let us consider, for example, the non-linear logics mentioned at the end of Section 5.5, whose connectives are implicitly defined with the help of an ordering on the truth values. These logics are good candidates for logics being used in independence proofs (Gabbay, 1991) and we could well imagine a mechanical proof assistant for independence proofs based on a versatile many-valued theorem prover. The user would specify designated truth values, truth value order, and definitions of connectives, while the machine would automatically compute the truth tables, instantiate a generic theorem prover with them, and check whether they constitute an independence proof for a given axiom system. With such a tool, one's intuition could be checked quickly and tedious computations could be delegated to a machine.

7.2.2 *Relevance logic: A special case*

The purpose of the remaining subsections is mainly to show that a number of developments in many-valued logics can in fact be regarded as special cases of the sets-as-signs framework developed in the preceding chapters.

Belnap (1977) defined a four-valued propositional logic which corresponds to the axiom system of the first-order entailment logic in Anderson and Belnap Jr (1975). An attractive feature of this logic is that it can deal with contradictory information. This has been exploited, for example, in Fitting (1989) to model the semantics of distributed logic programs. In the view of Ginsberg (1986a) the ordering induced on the set of truth values by taking $\wedge$ as infimum and $\vee$ as supremum constitutes the simplest non-trivial bilattice. Belnap's logic comprises the connectives $\wedge, \vee, \rightarrow, \neg$ which are defined according to Table 7.1. As the set of truth values we take $N = \{\mathsf{None}, \mathsf{F}, \mathsf{T}, \mathsf{Both}\}$. The choice of terminology becomes clear if we note that the connectives can alternatively be defined as

$$
\begin{aligned}
i \wedge j &= \inf(i, j) \\
i \vee j &= \sup(i, j) \\
i \rightarrow j &= \begin{cases} \mathsf{T} & i \preccurlyeq j \\ \mathsf{F} & \text{otherwise} \end{cases}
\end{aligned}
$$

where $\inf, \sup, \preccurlyeq$ are induced by the lattice in Figure 7.1. The truth value Both is understood to support truth and falsehood while None supports

Table 7.1   *Truth tables of relevance logic*

∧	None	F	T	Both
None	None	F	None	F
F	F	F	F	F
T	None	F	T	Both
Both	F	F	Both	Both

∨	None	F	T	Both
None	None	None	T	T
F	None	F	T	Both
T	T	T	T	T
Both	T	Both	T	Both

→	None	F	T	Both
None	T	F	T	F
F	T	T	T	T
T	F	F	T	F
Both	F	F	T	T

	¬
None	None
F	T
T	F
Both	Both

neither. The connective → occurs only on the top level and models a meta-level entailment relation.

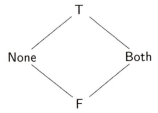

FIG. 7.1. The truth value lattice of relevance logic.

D'Agostino (1990) gave a simple signed tableau system for Belnap's logic which relied on signs $\{t, f, t^*, f^*\}$, which have the implicit meaning *supports truth, does not support truth, does not support falsehood,* and *supports falsehood.* D'Agostino proved the soundness and completeness of his system for Belnap's logic, that is, $\phi \to \psi$ holds iff there is a closed tableau for $\{t\ \phi, f\ \psi\}$.

If we take $\{t, f, t^*, f^*\}$ as abbreviations for

$$
\begin{aligned}
t &= \{\text{Both}, \text{T}\} \\
f &= \{\text{None}, \text{F}\} \\
t^* &= \{\text{None}, \text{T}\} \\
f^* &= \{\text{Both}, \text{F}\}
\end{aligned}
$$

work exactly with these signs, and use Lemma 5.4.1 from D'Agostino (1990), we obtain D'Agostino's tableau system *exactly* as a special case

from the systems introduced in Chapter 4, while soundness and completeness results follow from Theorems 4.14 and 4.21 without any further work
being required.

Hence, D'Agostino (1990, pp. 92–100) is just a special case of our technique. Note that we can also easily model other entailment relations, as
long as they are truth-functional.

If we observe that $t =\uparrow$ Both, $f =\downarrow$ None,... it becomes obvious that a
formulation of the rules in the spirit of Section 5.5 which can take advantage
of the lattice ordering could simplify the tableau system further.

7.2.3 *Suchoń's tableau system: another special case*

Probably the first tableau axiomatization of an n-valued logic was provided by Suchoń.* In Suchoń (1974) a tableau system for finitely valued
Łukasiewicz logics with truth value set $N = \{0, \frac{1}{n-1}, \ldots, 1\}$ is given. As in
Surma (1984), Carnielli (1987), D'Agostino (1990), Doherty (1991), Murray
and Rosenthal (1991b), and in our own work, the technical device is a
suitable modification of the sign language.

The sign language in Suchoń (1974) consists of the set

$$\{T_0, T_1, \ldots, T_n, F_0, F_1, \ldots, F_n\}$$

A branch constructed using the rules according to Suchoń is closed iff
it contains signed formulas $T_i\,\phi, F_j\,\phi$ with $i \leq j$. A formula ϕ is proved by
constructing a closed tableau with root $F_1\,\phi$.

Suchoń does not motivate or explain his construction, but merely gives a
formal soundness and completeness proof of his system based on a modified
definition of the Hintikka sets—consequently it is quite difficult to see how
the method works. Using the tools developed in Chapter 4 we can shed
some light on it and see that it is another special case of our sets-as-signs
notion. The correspondence is given below.

We regard T_i, F_j as abbreviations for

$$T_i = \left\{ j \mid j \geq \frac{n-i}{n-1}, j \in N \right\}$$
$$F_i = \left\{ j \mid j < \frac{n-i}{n-1}, j \in N \right\}$$

Suchoń works with the redundant signs $T_n = F_0 = N$. If we subtract
these we are left exactly with our regular signs, that is, $T_i = \boxed{\geq \frac{n-i}{n-1}}$ if
$i < n$ and $F_{i-1} = \boxed{\leq \frac{n-i}{n-1}}$ if $i > 1$. Hence, Suchoń's closure conditions
coincide with our own.

---

*Of course, before this there have been Gentzen style axiomatizations of many-valued
logics which are easily seen to be equivalent to tableaux; see Section 8.2.2.

A short inspection is sufficient to show that with this definition of T_i, F_j Suchoń's rules are *exactly* those yielded by Definition 4.4 modulo redundant formulas $T_n\,\phi, F_0\,\psi$ which do not carry any information.

Thus we see that Suchoń's system is just a special case of Chapter 4. While the sets-as-signs notion is central to our approach, however, it is only present implicitly in Suchoń's work.

7.2.4 *Improvement of an* **S5***-implementation by Caferra and Zabel*

Caferra and Zabel (1990) use a many-valued theorem prover based on the propositional part of Carnielli's (1987) work to implement a theorem prover for some propositional modal logics. The key observation (Hughes and Cresswell, 1984; McRobbie *et al.*, 1988) is that for any formula ϕ in certain modal logics it is possible to give a bound m_ϕ on the size of a potential Kripke model depending on the size of the formula. In order to refute satisfiability of a formula it is sufficient to refute all models with at most m_ϕ possible worlds. If one encodes an m-world model with 2^m truth values, one can emulate modal logic with many-valued logic whenever a suitably flexible many-valued theorem prover is available.

If we denote vectors of classical truth values as binary numbers $\langle k_1, \ldots, k_m \rangle$, the many-valued signed formula $\langle k_1, \ldots, k_m \rangle\,\phi$ is associated with a Kripke model where ϕ holds in world j iff $k_j = 1$. With this translation at hand it is not difficult to specify many-valued connectives that characterize modal operators in m-world models.

Although Caferra and Zabel identify branches that are identical up to a permutation of the worlds in the signs, branching may still be excessively high for more complex formulas.*

Here our framework using sets-as-signs can save much of the branching that cannot be saved by identifying equivalent states. In addition, two of the approaches described in Chapter 6 may prove valuable. First, one could order the truth value set in such a way as to reflect the modal structure; the signs $\boxed{\geq i}$ in this context could, for example, mean 'all truth values corresponding to states that can be reached from state i'. This is not so interesting for the modal logic S5 (Caferra and Zabel's only example), but, rather, for modal logics with weaker accessibility relations. Second, incorporating constraints might prove extremely interesting in this setup. This time the constraints would not be arithmetic inequalities, but would rather reflect accessibility constraints. Some connections with the work of Ohlbach (1989) show up here; however, it remains to be seen whether a tableau-based approach can offer any advantage over his resolution-based approach.

---

*It is probably best to iteratively increase the number of worlds starting with 1, since the bound m_ϕ coming from techniques as in Hughes and Cresswell (1984) is usually not very sharp.

7.3 Applications of a practical nature

7.3.1 *Interval arithmetic*

Interval arithmetic is an attempt to extend the usual operators of arithmetic from natural (integer, rational, real)* numbers to *intervals* of such numbers. An interval is understood as a number that is not precisely known, or, more accurately, that is known up to a certain degree of error. Possible applications of interval arithmetic are the investigation of rounding errors in arithmetical computations or the development of new proof techniques in number theory.

The main problem that occurs is the definition of appropriate order relations on intervals. For example, should $[1, 3] \leq [2, 4]$ hold or not? A cautious strategy would suggest that $[a, b] \leq [c, d]$ iff $b \leq c$. This leads to very counter-intuitive results (such a relation would not be reflexive). On the other hand, when $[a, b] \leq [c, d]$ iff $a \leq d$ transitivity is lost. Similar problems occur for equality.

For these reasons many-valued models were considered relatively early on by interval mathematicians, but have not been pursued further until recently. In Gerberding (1991) a three-valued semantic for interval arithmetic is defined and interval analogues of the Peano axioms are given that characterize the three-valued model categorically.

An interesting feature of the logic used is that it has two designated truth values. Our example from above $[1, 3] \leq [2, 4]$ would be assigned the intermediate truth value, denoting a kind of weak affirmation of the ordering relation. Gerberding (1991) demonstrates that for his choice of truth value designation and definition of $\leq, =, +, *, 0$ a natural behaviour of the operators is obtained.

The first-order part of the logic has been axiomatized as a tableau system using our sets-as-signs rules and several properties of $\leq$ and $=$ could be proved automatically, including, for example, antisymmetry and transitivity of $\leq$, validity of axioms of the Hilbert system, and commutativity of $+$.

7.3.2 *Hardware verification*

In Hardware Verification, several potential application areas for a many-valued theorem prover can be found:

- verification of genuinely many-valued circuits,
- modelling dynamic hazards by introducing pseudo states to find overlapping regions of competing signals (race detection) (Brzozowski and Seger, 1991),
- test pattern generation by propagation of *undefined* or *error* values (Cho and Bryant, 1989),

---

*For the sake of simplicity we consider only natural numbers in what follows.

- verifying the implementation of gates on the basis of switch level models (Hayes, 1982; Hayes, 1986).

For various reasons the last alternative seems to be the most attractive. Switch level models have been introduced by Hayes (1982) and Bryant (1984) and allow the formal description of circuits at quite a detailed level. For example, wire capacities, different voltage levels, and design techniques such as CMOS or NMOS are accounted for, usually however, on a symbolic level without specifying concrete values. Also, the actual circuit geometry and layout is not an issue on this level. The target of the specification is a single gate, for example an inverter or a NAND gate.

The 'grain size' of a switch level model can be varied; in a first implementation one would, for example, exclude handling of wire capacities in order to keep the number of required truth values small.

In a structured approach to hardware verification ranging from physical dimensions of parts and wires to a high-level description of module functions (which in turn might be the lowest level of software verification) there would typically be one layer beneath the switch level and two or more layers above the gate level. Each layer is specified by means of the layer directly beneath it; on each level specifications can be completely formal, although the tools may be quite different.

It turns out that switch level specifications can be described naturally in many-valued logic. Various voltage levels in different situations are realized with different truth values (Hayes, 1986). Transistors are switches and their propagation of voltage levels can be modelled by suitable logical connectives. As opposed to other approaches which provide only simulation tools for switch level specifications (Bryant, 1987), the many-valued logic approach allows automatic *verification*.

In Kernig (1992) and Hähnle and Kernig (1993) a case study for a particular switch level model for CMOS circuits has been carried out. The many-valued theorem prover $_3\mathcal{T^AP}$ (see Section 6.4.2) is used to verify switch level specifications of small gates automatically. The simplest logic that was used has seven truth values. More sophisticated models would have more truth values. Although we cannot go into the details here, some points about the logics used are noteworthy:

- None of the many-valued logics used has occurred before in the literature outside the field of switch-level theory.
- For some connectives the easiest way to provide a semantic is via certain lattices on the truth values. The truth tables of other connectives are determined by the physical properties of transistors and the kind of phenomena that are to be modelled; there is no mathematically natural description.
- For dealing with *sequential* circuits (that is, circuits with feedback, in contrast to *combinational* circuits) it would be advantageous to have

temporal operators.

Thus we need a theorem prover for many-valued logics which is not only suitable for a wide range of logics, but which can also be easily adapted and is able to handle logics with a relatively large number of truth values. In the light of the lattice-based semantics of the logics involved it is particularly interesting to work out the details from Section 5.5, although the sets-assigns approach alone is satisfactory for a prototype.

Tableau-based approaches to temporal logic theorem proving already exist (Janssen, 1989; Wolper, 1981); using constraint techniques as in Section 6.2 they could possibly be made more efficient.

Other approaches to hardware verification on the switch level that involve many-valued logic are reported in Bryant and Seger (1991) and Filkorn *et al.* (1991). The first approach works by a reduction of a temporal multiple-valued logic to classical propositional logic. The second states many-valued deduction problems as unification problems in finite algebras. These are reduced to unification in certain Boolean algebras. Boolean unification is implemented using decision diagrams (see also Section 8.2.1).

We argue that genuinely many-valued reasoning techniques are potentially more efficient than reduction. Moreover, many-valued automated deduction is flexible and not restricted to any particular class of logics (for instance, the reduction from finite algebras to Boolean algebras works only for functionally complete logics).

8

A HISTORY OF MULTIPLE-VALUED THEOREM PROVING

Without any doubt it is a discovery of the first order, eclipsing everything done in the field of logical research in Poland
— Zbigniew Jordan, *on many-valued logic*

The literature on many-valued systems is ridiculously large, but I am sure it does not match in waste of time that on modalities
— Dana Scott, *on the same topic*

Introduction

Historically, most proof systems for many-valued logics were based on some version of the resolution rule; consequently, these will be reviewed together in a separate section. All but one of the remaining approaches are based upon extensions of more or less known proof procedures for classical logic. For the sake of completeness we have also included two very general frameworks for theorem proving which are not limited to many-valued logics. Two of the reviewed approaches (Sections 8.1.4.3 and 8.2.4) can deal with infinitely valued logics, although in a limited way. We treat only such proof systems extensively that were intended for automated theorem proving by their authors or that are obviously suitable for that purpose. In particular, we do not mention most of the vast literature on Hilbert-style or Gentzen-style axiomatizations. See Wolf (1977), Rescher (1969), Urquhart (1986), or Gottwald (1989), and the references contained therein, for that kind of proof systems.

In the following overview we had to translate and partly reinterpret the definitions and results in our notation to achieve a certain degree of homogeneity. Those definitions and theorems that bear a reference to one or more papers in their titles have been taken more or less directly from the specified work(s). Others that were scattered amongst various articles required some reinterpretation within the context of the framework used. The relevant and most accessible papers on each approach are listed at the beginning of the respective subsection.

8.1   Resolution-based systems

Common to most resolution-based approaches (see Stachniak's system for an exception) and common to all systems using a uniform inference rule (such as Dissolution or Plaisted's method) is the requirement that theorems have to be transformed into some kind of normal form before the deduction process can even start. For classical propositional and predicate logic, how to achieve various normal forms is well known and considerable research has been undertaken on algorithms that produce short normal forms; see, for example, Boy de la Tour (1990). In many-valued logics, however, normal forms often do not even exist, let alone they can be computed efficiently.

Based on the sets-as-signs notion propagated in this book, some results regarding normal forms for many-valued logics are now available (see Hähnle (1993b) and Murray and Rosenthal (1993b); and Sections 6.3.4 and 6.3.3). All the earlier approaches presented in the following do not use sets-as-signs and this, together with the fact that most multiple-valued logics lack the existence of *unsigned* normal forms,* impose a severe restriction on the applicability of resolution-based systems, at least on those which employ a clausal form of the resolution rule.

8.1.1   *Morgan's resolution-based system*

Perhaps the earliest reference to a many-valued deduction system, purposely designed for automated theorem proving and that has also been actually implemented, is Morgan (1976). The primary motivation for Morgan was to have a finite approximation to fuzzy logic systems (for an overview see, for example, Zadeh (1988)), but in a sense it turns out to be more general.

Morgan considers for each finite number n of truth values the class $\mathcal{L}_M^n$ (Section 2.3.1) of many-valued first-order logics that contain, besides conjunction, disjunction, negation, universal, and existential quantification with the usual multiple-valued generalized semantics, n unary connectives $J_0, \ldots, J_1$ which intuitively assert that the argument of J_i has truth value i.

In Section 2.3.1 we mentioned that for each n the propositional part of $\mathcal{L}_M^n$ is functionally complete. Vice versa, any n-valued logic, whose propositional part is functionally complete could potentially be handled by Morgan's approach. Even if the propositional part of a logic is not functionally complete, it could be handled by considering a suitable functionally complete conservative extension, the queries to which are restricted to formulas that can be expressed in the first system.

This generality, however, is of limited practical use because of the fixed set of connectives: although arbitrary many-valued propositional connec-

---

*For an exception, which also shows how difficult the problem may become, see Mundici (1991; 1993).

tives are definable in $\mathcal{L}_M^n$, the actual definition can be awkwardly long. In addition, it is a non-trivial task to find a minimal representation for a many-valued propositional function over a given set of operators—in fact the problem is only solved satisfactorily in special cases. See Section 4.5 for more on this issue.

Morgan's two main results reflect the two main problems which resolution-based and other similar approaches to many-valued theorem proving must solve:

(i) to give a sound and complete transformation algorithm that computes the normal form of a given formula such that it can serve as input to

(ii) a generalized inference rule that must be proved sound and complete, usually by adapting a standard proof.

Definition 8.1. (J-CNF (Morgan, 1976)) *A* $\mathbf{L}_M^n$*-formula is said to be in* **J-CNF** *iff it is in CNF and all literals occurring in it are of the form* $J_i(p)$*, for some* $i \in N$ *and* p *atomic.*

Morgan proves a sequence of theorems which can be summarized as follows.

Theorem 8.2. (Normal form (Morgan, 1976)) *To each closed* $\mathbf{L}_M^n$*-formula* ϕ*, there corresponds a closed formula* ψ *in* $(\mathbf{L}_M^n)^{\mathbf{sko}}$ *which is in universal prenex J-CNF such that* ϕ *is satisfiable iff* ψ *is satisfiable.*

Note, however, that we cannot replace 'satisfiable' in the theorem above by 'has truth value i in the same structure', since the transformations given in the proof preserve satisfiability, but may actually alter the truth value. We can formalize the difference between both notions as follows.

Definition 8.3. (Refutational, logical equivalence) *Let* $\mathcal{L}$ *denote a many-valued first-order logic. Two closed formulas* $\phi, \psi \in \mathbf{L}$ *are called* **logically equivalent** *iff* $v(\phi) = v(\psi)$ *in all structures for* $\mathcal{L}$*. They are called* **refutationally equivalent** *iff* $(v(\phi) \in D$ *iff* $v(\psi) \in D)$ *in all structures for* $\mathcal{L}$*.*

Since in refutation systems we are only interested in the coarser notion of refutational equivalence, Theorem 8.2 is sufficient. In many-valued resolution approaches this property is often exploited in order to obtain simpler systems, as we will see in the following sections.

We note for later use that both equivalence relations are congruences on $\mathbf{L}$.

Proposition 8.4. *Let* ϕ, ψ *be closed* $\mathcal{L}$*-formulas and* **Abbr** *a finite set of abbreviations in* $\mathbf{L}$*.*

1. *If for any abbreviation* $\psi[\phi_1, \ldots, \phi_n]_{def} = \theta[\phi_1, \ldots, \phi_n] \in \mathbf{Abbr}$ *and substitution* $\sigma : \mathbf{fs} \to L$ $\psi[\phi_1, \ldots, \phi_n]\sigma$ *and* $\theta[\phi_1, \ldots, \phi_n]\sigma$ *are logically*

(refutationally) equivalent, then for any $\phi', \psi' \in \mathbf{L}$ ϕ' and ψ' are logically (refutationally) equivalent whenever ϕ' and ψ' are syntactically equivalent.

2. *If ϕ and ψ are logically equivalent then ϕ and ψ are also refutationally equivalent.*

Proof. 1. Exactly as in the classical case.

2. Immediate from the definition.

Using the weak connectives of $\mathcal{L}_{M+}^n$ it is possible to give an object level formulation of Theorem 8.2:

Corollary. To each closed $\mathbf{L}_M^n$-formula ϕ, there corresponds a closed formula ψ in $(\mathbf{L}_M^n)^{\mathrm{sko}}$ which is in universal prenex J-CNF such that $\vDash_{\mathcal{L}_{M+}^n} \phi \equiv \psi$ holds.

Formulas in universal prenex J-CNF are being processed further with a straightforward modification of the two-valued resolution rule. The basic idea is as follows. In J-CNF clauses only literals of the form $J_i(p)$ for some $i \in N$ and atomic p can occur. A disjunction of such literals, say $J_{i_1}(p) \vee \cdots \vee J_{i_k}(p)$, says informally that p has to take on a truth value from the set $\{i_1, \ldots, i_k\}$. Now assume that we have two clauses of the form $C_1 \vee J_{i_1}(p) \vee \cdots \vee J_{i_k}(p) \vee D_1$ and $C_2 \vee J_{j_1}(p) \vee \cdots \vee J_{j_{k'}}(p) \vee D_2$. Then the sets $\{i_1, \ldots, i_k\}$ and $\{j_1, \ldots, j_{k'}\}$ are the truth values that p can take on in the first and the second clause respectively. If these sets have an empty intersection then the factors $J_{i_1}(p) \vee \cdots \vee J_{i_k}(p)$ and $J_{j_1}(p) \vee \cdots \vee J_{j_{k'}}(p)$ obviously cannot be satisfied simultaneously and can thus be omitted from the analysis of both clauses.

Definition 8.5. (Resolution principle (Morgan, 1976)) *Let C_1 and C_2 be the sets of literals in two clauses from the matrix of a J-CNF formula in $\mathcal{L}_M^n$. Without loss of generality assume that they have no free variables in common. Let $K_1 = \{J_{i_{11}}(p_{11}), \ldots, J_{i_{1r_1}}(p_{1r_1})\}$ and $K_2 = \{J_{i_{21}}(p_{21}), \ldots, J_{i_{2r_2}}(p_{2r_2})\}$ be subsets of C_1 and C_2 respectively. If*

- $\{p_{11}, \ldots, p_{1r_1}, p_{21}, \ldots, p_{2r_2}\}$ *are unifiable with most general unifier σ,*
- $\{i_{11}, \ldots, i_{1r_1}\} \cap \{i_{21}, \ldots, i_{2r_2}\} = \emptyset$,

then a clause with the set of literals $(C_1 - K_1)\sigma \cup (C_2 - K_2)\sigma$ is called an **mv-resolvent** *of C_1 and C_2.*

Remark 8.6. 1. *This rule is more general than necessary. It is easy to see that each general mv-resolution step can always be substituted by a finite sequence of mv-resolution steps where the sets K_1 and K_2 are singletons. On the other hand, one could extend the rule to more than two input clauses to a kind of many-valued hyperresolution.*

2. *The sets K_1 and K_2 in the above definition are not required to be maximal. On the contrary, to achieve completeness all possible mv-resolvents from a set of clauses must be generated within a finite number of steps. Obviously, this enlarges the search space dramatically.*

The notion of a resolution proof from here on is completely standard. With these definitions at hand it is not too difficult to prove a generalized form of Herbrand's theorem and hence the soundness and completeness of the mv-resolution rule for $\mathcal{L}_M^n$.

Theorem 8.7. (Soundness, completeness (Morgan, 1976)) *Let ϕ be some $\mathbf{L}_M^n$-formula and ψ a universal prenex J-CNF of ϕ. Let E be the set of clauses corresponding to the matrix of ψ. Then ϕ is unsatisfiable iff $\square$ can be obtained by a finite number of mv-resolution steps from E.*

Let us conclude this section with an example of a derivation in Morgan's system.

Assume that we want to prove theoremhood of a formula from $\mathbf{L}_{M+}^n$ within the resolution system for $\mathcal{L}_M^n$. Since $\mathcal{L}_M^n$ is functionally complete, we can express $\sim$, $\supset$, $\equiv$ and $\cong$ with the help of the remaining operators. One possibility would be to use the following abbreviations:

$$\sim \phi \quad _{def} = \quad \neg(J_{\frac{n-d}{n-1}}(\phi) \vee \cdots \vee J_1(\phi))$$

$$\phi \supset \psi \quad _{def} = \quad \sim \phi \vee \psi \tag{8.1}$$

$$\phi \equiv \psi \quad _{def} = \quad (\phi \supset \psi) \wedge (\psi \supset \phi) \tag{8.2}$$

$$\phi \cong \psi \quad _{def} = \quad (J_0(\phi) \wedge J_0(\psi)) \vee \cdots \vee (J_1(\phi) \wedge J_1(\psi))$$

Remark 8.8. *It is easy to check that for all instances in the formula schemata above, both sides are logically equivalent. Making use of Proposition 8.4 we will thus write $\equiv$ instead of $\Leftrightarrow$ from now on.*

Example 8.9. *We give a proof that the formula $\neg p \supset (\sim p \wedge \neg p)$ is a tautology in $\mathcal{L}_M^3$, when 1 is the only designated truth value.*

It is easy to see that for this purpose it is sufficient to prove that the weakly negated formula is unsatisfiable, in other words, the truth values 0 and $\frac{1}{2}$ are not taken on in any structure.

We begin by expanding the definitions of $\sim$ and $\supset$:

$$\sim (\neg p \supset (\sim p \wedge \neg p)) \equiv \neg J_1(\neg J_1 \neg p \vee (\neg J_1 p \wedge \neg p))$$

Next, the unary operators are pushed in front of the atoms and the conjunctions are pushed to the top. A lot of derived simplification rules are needed for this purpose. To avoid astronomical growth of the resulting clause set, subsumption checks also have to be made after each transformation step. After several hundred transformation steps where intermediate formulas such as

$(J_0 \neg J_0 p \wedge J_0 \neg J_2 p) \vee (J_1 \neg J_0 p \wedge J_1 \neg J_2 p) \vee (J_1 \neg J_0 p \wedge J_0 \neg J_2 p) \vee$
$(J_0 \neg J_0 p \wedge J_1 \neg J_2 p) \vee (J_0 \neg J_0 p \wedge J_0 \neg p) \vee (J_1 \neg J_0 p \wedge J_1 \neg p) \vee$
$(J_1 \neg J_0 p \wedge J_0 \neg p) \vee (J_0 \neg J_0 p \wedge J_1 \neg p)$

occur, the final result is

   (1) $J_0 p$
   (2) $J_0 p \vee J_1 p$
   (3) $J_2 p \vee J_0 p$
   (4) $J_2 p \vee J_1 p$

A refutation of this clause set is trivial:

   (5) $\Box$ *from (1) and (4)*

Morgan's system can be used, at least theoretically, to conduct resolution proofs in a wide variety of many-valued logics. Its practical use, however, is limited due to the fact that the definition of complex new propositional operators can become tedious and long, since it has to be done solely on the basis of the J-connectives and, therefore, normal form computation soon becomes intractable.

Recently, a general treatment of resolution for many-valued logics based on signed clauses has been published (Baaz and Fermüller, 1992). The resolution rule used in that paper is the simplified version of Morgan's rule (see Remark 8.6, part 1). The propositional CNF transformation rules are those from Section 6.3.4; however, only singleton sets as signs and no structure preserving clause form are employed. On the other hand, a full treatment of distribution quantifiers and many-valued versions of some classical resolution refinements are included.

8.1.2 *Schmitt's resolution-based system*

The resolution-based proof system given by Schmitt (1986) for a certain three-valued logic that we will henceforth call $\mathcal{L}_3$ is very much in the same spirit as Morgan's approach reviewed in Section 8.1.1. The logic $\mathcal{L}_3$ was introduced by Fenstad *et al.* (1985) in connection with natural language processing. The motivation for Schmitt was to provide a proof system for $\mathcal{L}_3$ that could be implemented reasonably efficiently for use within a natural language dialogue system. We will see that the drawback of Morgan's approach mentioned at the end of Section 8.1.1 does not count so much in the case of $\mathcal{L}_3$. First, the language, the number of truth values, and the subset of designated truth values is fixed and, second, the choice of connectives is such that the system obtained is relatively small.

Definition 8.10. *($\mathcal{L}_3$) Let $\mathcal{L}_3$ be a three-valued first-order logic according to the following specifications:*

  • *The associated propositional language is given by $(L_3, \wedge, \vee, \neg, \sim)$ with similarity type $\langle 2, 2, 1, 1 \rangle$.*

- *The matrix corresponding to the associated propositional language $\mathbf{L}_3$ is $(\mathbf{3}, 2, \wedge, \vee, \neg, \sim)$, where the operators are defined as in Tables 2.1 and 2.2.*

We note that $\mathcal{L}_3$ is not functionally complete,[*] but weak implication and equivalence as defined in Table 2.2 can be expressed as in (8.1) and (8.2).

Definition 8.11. (S-CNF) *A $\mathbf{L}_3$-formula is said to be in **S-CNF** iff it is in CNF and all literals occurring in it are of the form $p, \neg p, \sim p$ or $\sim \neg p$ where p is atomic.*

Theorem 8.12. (Normal form (Schmitt, 1986)) *To each closed $\mathbf{L}_3$-formula ϕ, there corresponds a closed universal prenex S-CNF formula ψ in $(\mathbf{L}_3)^{\text{sko}}$, such that the expression $\phi \equiv \psi$ is a first-order tautology.*

One observes that no additional connectives have to be introduced in order to achieve S-CNF normal form with only four types of literals. The reason is that any unary connective that is composed of zero or more applications of $\neg$ and $\sim$ is refutationally (though not logically) equivalent to one of the four literals occurring in S-CNF. Of course this also greatly simplifies the resolution rule.

In the following definition let us denote a literal with $s(p)$, where s is one of $\{\epsilon, \neg, \sim, \sim \neg\}$.[†]

Definition 8.13. (Three-valued resolution (Schmitt, 1986))   *Let C_1 and C_2 be the sets of literals in two clauses from the matrix of an S-CNF formula in $\mathcal{L}_3$. Without loss of generality assume that they have no free variables in common. Let $K_1 = \{s_{11}(p_{11}), \ldots, s_{1r_1}(p_{1r_1})\}$ and $K_2 = \{s_{21}(p_{21}), \ldots, s_{2r_2}(p_{2r_2})\}$ be subsets of C_1 and C_2 respectively. If*

- *$\{p_{11}, \ldots, p_{1r_1}, p_{21}, \ldots, p_{2r_2}\}$ are unifiable with a most general unifier σ, and*
- *$\{s_{11}, \ldots, s_{1r_1}\} \subseteq \{\epsilon\}$ and $\{s_{21}, \ldots, s_{2r_2}\} \subseteq \{\neg, \sim\}$, or $\{s_{11}, \ldots, s_{1r_1}\} \subseteq \{\neg\}$ and $\{s_{21}, \ldots, s_{2r_2}\} \subseteq \{\epsilon, \sim \neg\}$,*

*then a clause with the set of literals $(C_1 - K_1)\sigma \cup (C_2 - K_2)\sigma$ will be called a $\mathcal{L}_3$-**resolvent** of C_1 and C_2.*

The remarks we made on Morgan's resolution rule also apply here.

Theorem 8.14. (Soundness and completeness (Schmitt, 1986)) *Let ϕ be a $\mathbf{L}_3$-formula and ψ a universal prenex S-CNF of ϕ. Let E be the set of clauses corresponding to the matrix of ψ. Then ϕ is unsatisfiable iff $\square$ can be obtained by a finite number of $\mathcal{L}_3$-resolution steps from E.*

Let us illustrate this with the example from Section 8.1.1.

---

[*]The proof of this fact can be found in Schmitt (1990).
[†]Let ϵ denote the empty string.

Example 8.15. *We give a proof that the formula $\neg p \supset (\sim p \wedge \neg p)$ is a tautology in $\mathcal{L}_3$. As before, we start by negating the formula to be proved and expand the definition of $\supset$.*

$$\sim [\neg p \supset (\sim p \wedge \neg p)]$$
$$\equiv \;\; \sim [\sim \neg p \vee (\sim p \wedge \neg p)]$$

Computing S-CNF yields:

$$\equiv \;\; \sim\sim \neg p \wedge (\sim\sim p\vee \sim \neg p)$$
$$\equiv \;\; \neg p \wedge (p\vee \sim \neg p)$$

This gives as the starting clause set:
    (1) $\neg p$
    (2) $p\vee \sim \neg p$
A refutation of this clause set is trivial:
    (3) $\square$ from (1) and (2)

It may not seem fair to take an example that contains mostly connectives that are built into the logic. The point we want to illustrate, however, is that functional completeness alone is not sufficient. If the required operators are present in the deduction system, much better results can be achieved.

The relative simplicity of Schmitt's system rests mainly on two facts that are worth remembering.

Remark 8.16.   1. *The unary connectives obey DeMorgan's laws with respect to conjunction and disjunction. Therefore, computation of the normal form follows more or less the classical case and is not too expensive.*

  2. *The fact that in a resolution system, refutationally equivalent as opposed to logically equivalent transformations are sufficient is used to minimize the required instances of the resolution rule.*

8.1.3 *Stachniak's resolution logics*

In a series of papers (Stachniak, 1988; Stachniak and O'Hearn, 1990; Stachniak, 1990a; Stachniak, 1990b; O'Hearn and Stachniak, 1992; Stachniak, 1991a; Stachniak, 1991c; Stachniak, 1991b; Stachniak, 1992) Stachniak developed a framework for many-valued resolution that in certain respects is more general than the previous systems.

To see how and why, let us first recall that the usual definition of logical consequence in classical logic can mostly be carried over to non-classical

logics in more than one way.* Thus, minimum conditions that consequence
relations should satisfy have been postulated and the theory of logical consequences is a field of research in its own respect, which has been pushed
forward mainly by Polish logicians. Consequence relations provide an alternative in giving semantics to a language and from a certain point of view
they definitely have advantages over matrix-based semantics.

Although Stachniak has outlined a method as to how his results can
be lifted to first-order logic (O'Hearn and Stachniak, 1992), to keep the
presentation concise we will consider only propositional logics in the present
section.

Definition 8.17. (Consequence relation (Tarski)) *Let* L *be a propositional language. A function* $C : 2^L \to 2^L$ *is called a* **consequence relation** *on* L *iff the following conditions are satisfied for all* $X, Y \subseteq L$:

   (Ref) $X \subseteq C(X)$
   (Mon)   *If* $X \subseteq Y$ *then* $C(X) \subseteq C(Y)$
   (Id) $C(C(X)) \subseteq C(X)$

Every propositional logic $\mathcal{L} = (L, A)$ naturally defines a consequence
relation C_A on L:

$$\phi \in C_A(X) \text{ iff for every } \mathcal{L}\text{-valuation } v(X) \subseteq D \text{ implies } v(\phi) \in D$$

Conversely, a well-known result by Łoś and Suszko (1958) states the restrictions under which a given consequence relation has an adequate (possibly infinite) matrix. But general matrices[†] are still too large a class for
resolution.

Definition 8.18. (Strongly finite consequence relation) *Let* L *be a
propositional language. A consequence relation* C *on* L *is called* **strongly
finite (SF)** *iff there is a finite class* $\mathcal{K}$ *of finite propositional matrices for*
L *such that* $C(X) = \bigcap_{K \in \mathcal{K}} C_K(X)$ *for all* $X \in L$.

Obviously, any finitely-valued logic determines a strongly finite consequence relation; just take $\mathcal{K}$ to be a singleton.

Up to this point we have only considered proof procedures that perform
tests on satisfiability—the problem as to whether a formula follows from
a set of formulas is harder and in non-classical logics cannot in general be
reduced to the tautology problem.[‡] Since SF logics are decidable (Wójcicki,

---

*Strictly speaking, already in classical logic several variations of logical consequence
are possible, but in a sense the usual one encompasses the others.

†In general matrices need not to be finite, of course. But as we are mostly concerned
with finitely-valued logics, we mean finite matrices as in Definition 2.7 if not stated
otherwise.

‡Note that a formula is a tautology iff it is in $C(\emptyset)$. Thus unsatisfiability is merely a
special case of deciding C.

1988), there is an effective procedure that assigns to each $X \in L$ and $\phi \in C(X)$ a X_ϕ such that $\phi \in C(X)$ iff $C(X_\phi) = L$ (iff X_ϕ is inconsistent). It is this set X_ϕ that is refuted by a resolution system in order to prove that $\phi \in C(X)$.

As mentioned earlier, many-valued logics, and thus SF logics, may lack the expressive power to allow normal forms. Not even a disjunctive operator needs to be present. One way to deal with this situation is to embody the many-valued version of a non-clausal resolution system as it has been presented for classical logic (Murray, 1982; Manna and Waldinger, 1986).

In classical logic, the non-clausal resolution rule* has the following form.

$$\frac{\begin{array}{c}\phi(p)\\ \psi(p)\end{array}}{\phi(p)\{p \leftarrow \mathbf{f}\} \vee \psi(p)\{p \leftarrow \mathbf{t}\}}$$

where $\{\mathbf{f}, \mathbf{t}\}$ are the constant operators that always yield *false* and *true* respectively.

In SF logics neither $\vee$ nor $\{\mathbf{f}, \mathbf{t}\}$ have to be present, or even definable; hence, some modifications must be made.

- The disjunction in the conclusion of the resolution rule is replaced by a multiple-conclusion rule.
- The set of constant operators that gives the substitution for the resolution variable in the conclusion is replaced by a finite set of formulas $Ver= \{v_0, \ldots, v_m\}$, whose elements are called *verifiers* (we will define the verifiers precisely in a moment. For the time being it is sufficient to view them as a kind of truth value set).

The resolution rule then has the following form.

Definition 8.19. (SF resolution rule (Stachniak)) *Let the (not necessarily distinct) formulas* $\{\phi_0(p), \ldots, \phi_m(p)\}$ *all contain a variable* p *not occurring in* $Ver= \{v_0, \ldots, v_m\}$. *Then the following multiple-conclusion rule is called a* **SF logic resolution rule:**

$$\frac{\begin{array}{c}\phi_0(p)\\ \vdots\\ \phi_m(p)\end{array}}{\phi_0(p)\{p \leftarrow v_0\} \mid \quad \cdots \quad \mid \phi_m(p)\{p \leftarrow v_m\}}$$

In order to complete the system, two additional types of rules are necessary:

---

*We write down the rules as $n \times m$ *expansion rules* in the sense of D'Agostino (1990). The rules have n premises, m conclusions and are read as follows: 'If all formulas in the premise are on the current branch of the proof tree, generate m new branches each containing a formula from the conclusion and append them to the current branch.'

- *Transformation rules* that simplify the formulas generated by the resolution rule. These have the form

$$\frac{\phi(p)\{p \leftarrow F(v_0, \ldots, v_k)\}}{\phi(p)\{p \leftarrow v\}}$$

where $\{v_0, \ldots, v_k, v\} \in Ver$, F is a $(k+1)$-ary connective and v has to be logically equivalent to $F(v_0, \ldots, v_k)$

- $\Box$-*rules* that terminate the deductive process. These are all the inconsistent subsets $\{v_{i_1}, \ldots, v_{i_k}\}$ of Ver and the rules have the form

$$\frac{\begin{matrix} v_{i_1} \\ \vdots \\ v_{i_k} \end{matrix}}{\Box}$$

where $\Box$ is a special symbol not occurring in the language.

Definition 8.20. (Refutation tree, resolution system)   *We call a tree with root ϕ constructed according to the resolution, transformation, and $\Box$-rules given above a* **refutation tree** *iff all paths through the tree contain the symbol $\Box$. A set of verifiers together with accordingly constructed resolution, transformation, and $\Box$-rules for a SF logic (relative to its defining class of matrices $\mathcal{K}$) is called a* **resolution system** *for that logic.*

The crucial step towards a sound and complete resolution system for a given SF logic is, of course, the selection of a suitable set of verifiers. In classical logic we could take, for example, $Ver = \{p \supset p, \neg(p \supset p)\}$, when **f** and **t** are not present. In SF logics matters are more complicated; however, roughly speaking we require that, for each formula, there is a verifier that is logically equivalent to it. To make this precise we have to introduce some more terminology.

If $C_\mathcal{K}$ is a consequence relation on **L** induced by $\mathcal{K}$ and $P = \{p_1, \ldots, p_k\} \subseteq L_0$ are pairwise distinct variables we define the restriction $C_\mathcal{K}^P$ of $C_\mathcal{K}$ to P as follows:

$$\phi \in C_\mathcal{K}^P(X) \text{ iff } \phi\sigma \in C_\mathcal{K}(X\sigma) \text{ for each substitution } \sigma : L_0 \to P$$

A result by Wójcicki (1988, p. 256ff.) about SF logics says that it is always sufficient to consider $C_\mathcal{K}^P$ for some finite P:

Proposition 8.21. *Given any finite class of finite matrices $\mathcal{K}$ we have that $C_\mathcal{K}^P = C_\mathcal{K}$ for some finite P.*

Since there are only finitely many valuations from P into $\mathcal{K}$, the sublanguage of **L** generated by P is divided into finitely many equivalence classes by $\cong$.* These can be effectively constructed and, indeed, for any formula, there must be one class whose members are logically equivalent to it. Of course, these equivalence classes (or, more precisely, suitably chosen members from these equivalence classes) are taken as the verifiers.

If *Res* is a resolution system based upon a set of Verifiers *Ver*, as described above, then we let $X \overset{Res}{\Rightarrow} \square$ be an abbreviation of the fact that X shares no variables with *Ver* and has a refutation tree constructed with *Res*.

Theorem 8.22. (Resolution for SF logics (Stachniak)) *Let C be a SF consequence relation on $\mathbf{L}$ and let Res be a resolution system based upon a set of verifiers Ver defined as above. Then for any $X \subseteq L$*

$$C(X) = L \text{ iff } X \overset{Res}{\Rightarrow} \square$$

In Stachniak (1990a) it is shown that the number of verifiers may be arbitrarily large for certain classes of logics. More precisely, if t is the number of propositional variables occurring in *Ver* and k is the size of the smallest matrix in $\mathcal{K}$ then $|Ver|$ is bound by k^{k^t}. For each $t > 0$, however, there is a logic whose minimal resolution system requires at least t variables. For the classical case the bound is 4, whereas we have seen that 2 is sufficient. So the resolution systems thus generated are not by any means minimal and although an *effective* algorithm exists (Stachniak, 1990a; Stachniak, 1991a) that computes minimal sets of verifiers, the *efficient* construction of minimal systems so far has been addressed only for certain special cases, such as functionally complete logics or Łukasiewicz logics (Stachniak, 1990a).

Various suggestions to make the size (i.e. the number of verifiers) of the required resolution proof systems smaller have been proposed. One suggestion is to approximate a 'large' resolution system with a finite set of 'small' resolution systems (Stachniak, 1990b; Stachniak, 1992; Stachniak, 1991c):

Consider a resolution system *Res* which is sound and complete for a SF consequence relation C (in the sense of Theorem 8.22). Intuitively, it is clear that there must be smaller resolution systems than *Res* if we drop the soundness requirement. When such a class of small systems is suitably chosen, one can approximate *Res* in the following way:

Definition 8.23. (Resolution approximation) *Let C be a SF consequence relation and Res a resolution system for it. Let $\mathcal{R}$ be a finite class of resolution systems that are complete (but not necessarily sound) with respect to C. We say that $\mathcal{R}$ approximates Res iff for all $X \in L$:*

$$X \overset{Res}{\Rightarrow} \square \text{ iff } X \overset{R}{\Rightarrow} \square \text{ for all } R \in \mathcal{R}$$

---

*Where $\phi \cong \psi$ iff $\phi \cong_{\mathbf{K}} \psi$ for all $\mathbf{K} \in \mathcal{K}$.

Finding a set of small resolution systems that approximate a given system is a highly non-trivial task, for which no automatic procedures are in sight. The method has been extended to first-order logics (Stachniak, 1992) and And/Or-networks of approximating resolution systems (Stachniak, 1991c).

Another point where efficiency can be gained is the condition which is imposed on the transformation rules. It is possible to relax logical equivalence of the premise and conclusion to mere refutational equivalence without destroying soundness of the resulting resolution system (Stachniak, 1991a). This reduces not only the required number of verifiers but also enlarges the class of logics for which resolution systems can be found. This is due to the fact that there are logics which do not have SF consequence relations but share the same class of satisfiable formulas with a SF logic. For example, classical logic and intuitionistic logic have the same finite sets of inconsistent formulas. The condition

$$\phi \text{ is a tautology iff } \neg\phi \text{ is not satisfiable}$$

is, however, only valid in classical logic, which relativizes the achievement.

Other improvements consist of adopting the set of support and polarity strategies (Stachniak and O'Hearn, 1990; O'Hearn and Stachniak, 1992) from the classical case (Murray, 1982; Manna and Waldinger, 1986).

Finally, one might focus attention on logics containing a kind of disjunctive operator.[*] The result is that the multiple-conclusion rules can be substituted by single-conclusion rules with $\vee$ in the conclusion, which reduces the search space considerably.

Example 8.24. *For three-valued Łukasiewicz logic (Section 2.3.2) Stachniak gives the following six verifiers as a basis for a minimal resolution system (including the optimization regarding refutational equivalence as discussed above):*

$$v_{1 \ def} = p \supset p, \quad v_{2 \ def} = p \vee \neg p, \quad v_{3 \ def} = v_2 \supset \neg v_2$$

$$v_{4 \ def} = \neg v_3, \quad v_{5 \ def} = \neg v_2, \quad v_{6 \ def} = \neg v_1$$

Let us consider the following $\mathcal{L}_3$-tautology:[†]

$$\phi(r, q) := q \supset (r \supset q)$$

We illustrate Stachniak's system by showing that its strong negation is unsatisfiable. The paper size limit forces us to concentrate on a single branch in the following proof tree.

---

[*]A matrix is called *disjunctive* iff it has a connective $\vee$ such that for all $i, j \in N$: $i \vee j \in D$ iff $i \in D$ or $j \in D$, cf. Wójcicki (1988).

[†]This is an axiom of the first Hilbert-style axiomatization of $\mathcal{L}_3$, given by Wajsberg.

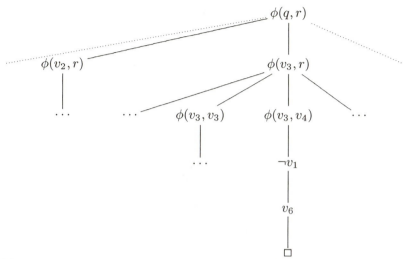

The upper two layers in the tree consist of applications of the resolution rule, while the next two layers are transformation rule applications followed by $\square$-rule applications.

Using disjunctive resolution rules linearizes the tree but does not change the size of the derivation.

To summarize, Stachniak's resolution-based approach to theorem proving in many-valued logics is of considerable flexibility and covers a wide scope of logics. Its main drawbacks are the size of the resulting systems, i.e. the branching factor of the rules due to the number of verifiers and the difficulty in finding small systems automatically (it is a non-trivial task already to design a small system for $\mathcal{L}_3$). Although some efforts have been made to overcome the first point, proofs that have been carried out by hand (see also the example above) indicate that an efficient implementation would be far from straightforward.

8.1.4 *Resolution-based systems for fuzzy logic*

In this section we will review very briefly several resolution-based approaches which all have in common that they are related to fuzzy logic* (Zadeh, 1965; Zadeh, 1988). As mentioned before fuzzy logic considered as a special case of multiple-valued logic was the motivation for Morgan's system, which we dealt with separately since it is of interest in its own right. Lee and Chang's system (Lee and Chang, 1971; Lee, 1972) was the first attempt at a many-valued version of the resolution rule. While the

---

*Actually there are two different notions of fuzzy logic around. The first, dealing with fuzzy logic as *local logic* is not considered here or in the cited papers. In theorem proving the second notion is always addressed, where a formula has a probabilistic truth value in the real interval $[0, 1]$.

work of Di Zenzo (Di Zenzo, 1988) can be seen as a direct descendant, the system of Orłowska (1978) is motivated by the $\omega+1$-valued logics that were originally introduced by Rasiowa (1973) in the context of program verification. Since both logics are based on Post algebras and their connectives are closely related it seems justified to present them together.

8.1.4.1 *Lee and Chang's resolution-based system*   Consider the first-order language **L** whose associated propositional language is defined by the free algebra $(L, \vee, \wedge, \neg)$ with the usual arities. The truth value set of the matrix is the unit interval $[0, 1]$ of the reals, whereas the matrix operations corresponding to the connectives are the same as in Definition 2.28. Note, however, that for the semantics of quantified formulas in Definition 2.19, min and max must be replaced with inf and sup respectively. The designated truth values are the closed interval $[0.5, 1]$.

As an additional notion Lee and Chang introduce the following definition.

Definition 8.25. (Falsified, unsatisfiable (Lee and Chang, 1971))
*Let ϕ be a sentence from L. We say that ϕ is **falsified** by a first-order structure **M** iff $v(\phi) \leq 0.5$ in **M**; it is **unsatisfiable** iff it is falsified under every structure.*

Using this definition it is possible that a formula is satisfied and falsified by the same structure. If we think without loss of generality of L-clauses as fully universally quantified, Lee's main result (1972) can be stated as follows.

Theorem 8.26. (Characterization of fuzzy logic (Lee, 1972))   *A set S of clauses is unsatisfiable in fuzzy logic iff it is classically unsatisfiable.*

In other words, under the present definition of unsatisfiability and a straightforward extension of two-valued semantics, no new theorems can be proved. The situation becomes different, of course, when one is interested in the actual truth value (probability) that an inferred formula takes on under a certain interpretation. Also, the fundamental property of classical consequence, namely that $v(\phi) \leq v(\psi)$ whenever ψ follows from ϕ, does not hold in fuzzy logic if ϕ can be falsified. Nevertheless, one would expect from a logic which claims to be a model for fuzzy reasoning a richer syntax that, for example, allows the probability of certain inference patterns to be specified generally, without assuming a specific interpretation. The examples in the next two subsections can count as witnesses for that line of development. A more recent paper is Weigert *at al.* (1993), which shows, however, few new aspects and is therefore not discussed in detail.

8.1.4.2 *Extended Post logics*   Di Zenzo (1988) considers an extension of Post logics (Section 2.3.3) which is enriched by an additional pair of disjunction and conjunction operators $\underset{m}{\vee}, \underset{m}{\wedge}$ for each *threshold value* $m \in \mathbf{n}$ and a negation operator $\neg$, formally.

Definition 8.27. (Extended Post logic (Di Zenzo, 1988)) *We call a first-order logic* $\mathcal{L}_{EP}^n$ *an* n**-valued extended Post logic,** *whenever*

- *its associated propositional language* $\mathbf{L}_{EP}^n$ *is given through the algebra*

$$(L_{EP}^n, \underset{0}{\wedge}, \underset{0}{\vee}, \ldots, \underset{1}{\wedge}, \underset{1}{\vee}, \sigma, \neg)$$

  with similarity type $\langle 2, 2, \ldots, 2, 2, 1, 1 \rangle$,
- *the definition of the corresponding matrix operations for* σ *and* $\neg$ *is as before, while for all* $i, j, m \in \mathbf{n}$:

  * $i \underset{m}{\wedge} j = \begin{cases} \max\{i, j\} & \text{if } i, j < m \text{ or } i, j \geq m \\ \min\{i, j\} & \text{otherwise} \end{cases}$

  * $i \underset{m}{\vee} j = \begin{cases} \min\{i, j\} & \text{if } i, j < m \text{ or } i, j \geq m \\ \max\{i, j\} & \text{otherwise,} \end{cases}$

- *designated truth values and thus satisfiability are not specified yet,*
- *instead of* $\forall, \exists$ *the first-order language contains* $2n$ *quantifiers* $\underset{m}{\forall}, \underset{m}{\exists}$, *which are defined by modifying Definition 2.19 in the following way:*

  * $v_\beta((\underset{m}{\forall} y)\phi) = \min\{\sup\{v_{\beta_y^u}(\phi) | v_{\beta_y^u}(\phi) \geq m, u \in U\},$
  $$\sup\{v_{\beta_y^u}(\phi) | v_{\beta_y^u}(\phi) < m, u \in U\}\},$$

  * $v_\beta((\underset{m}{\exists} y)\phi) = \max\{\inf\{v_{\beta_y^u}(\phi) | v_{\beta_y^u}(\phi) \geq m, u \in U\},$
  $$\inf\{v_{\beta_y^u}(\phi) | v_{\beta_y^u}(\phi) < m, u \in U\}\}.$$

Remark 8.28. *Obviously,* $\underset{0}{\wedge}$ *is identical to* $\vee$ *and the same is true for* $\underset{0}{\vee}$ *and* $\wedge$ *as well as for* $\underset{0}{\forall}$ *and* $\exists$, $\underset{0}{\exists}$ *and* $\forall$.

Since $\mathcal{L}_P^n$ is already functionally complete, in a certain sense all these additional operators are superfluous. They do allow, however, fuzzy relationships to be expressed more concisely than the ordinary system; moreover, they have interesting algebraic properties:

Let $\mathcal{L}_{EP}^{(n,m)}$ be the logic that is yielded when $\mathcal{L}_{EP}^n$ is restricted to contain only the operators $\underset{m}{\wedge}, \underset{m}{\vee}, \underset{1-m}{\wedge}, \underset{1-m}{\vee}, \neg$, and the quantifiers $\underset{m}{\forall}, \underset{m}{\exists}$.

It is easy to see that for given n all of the $\mathcal{L}_{EP}^{(n,m)}$ are isomorphic to each other and the subsystem for $m = 0$ is the n-valued restriction of Lee and Chang's system. Lee and Chang's results carry over to $\mathcal{L}_{EP}^{(n,m)}$ with suitable definitions of falsifiability and satisfiability. One can interpret m as the degree of confidence that describes the reliability of the data. Consequently, a derivation in $\mathcal{L}_{EP}^n$ consists of a sequence of resolution steps carried out in different subsystems $\mathcal{L}_{EP}^{(n,m)}$ that are appropriately chosen according to the reliability of the respective input clauses.

In conclusion, while from the standpoint of uninterpreted reasoning,[*] extended Post logics are not very interesting, one could very well imagine an application within a knowledge-based system relying on given probabilistic data.

8.1.4.3 ω+1-*valued Post logics*   In the early 1970s several infinitely-valued generalizations of Post logics have been investigated, mainly by Polish logicians. The primary motivation was to develop calculi for describing the logic of programs, and their interest seems to have been mainly in the algebraic properties and Hilbert-style proof theory of generalized Post algebras, but as Post algebras these logics are also potentially interesting for fuzzy applications. A step towards automation is Orłowska's resolution system (1978), see also Orłowska (1980, pp. 338–340), for a $\omega + 1$-valued logic from Rasiowa (1973).

Note that in the following definition for obvious reasons we do not take a subset of $[0, 1]$, but, rather, the chain $\omega + 1$ i.e. $0 < 1 < \cdots < \omega$ as the set of truth values.

Definition 8.29. (Generalized Post language) *We say that* **L** *is a generalized Post language iff* **L** *is a first-order language whose associated propositional language is defined by the algebra*

$$(L, \wedge, \vee, \supset, \neg, (D_i)_{1 \leq i < \omega}, (e_i)_{1 \leq i \leq \omega})$$

with similarity type $\langle 2, 2, 2, 1, 1^\omega, 0^{\omega+1} \rangle$.

In general, the semantics of **L** is given by *Generalized Post Algebras*, but if we assume the truth value set to be a chain with $\omega+1$ elements these algebras are all isomorphic (Rasiowa, 1973) and we can use the matrix:

$$(\omega + 1, \omega, \wedge, \vee, \supset, \neg, (D_i)_{1 \leq i < \omega}, \omega + 1)$$

where the operators are defined as follows:

- $D_i(j) = \begin{cases} \omega & i \leq j \\ 0 & i > j \end{cases}$

- $i \supset j = \begin{cases} \omega & i \leq j \\ j & i > j \end{cases}$

- $\neg i = \begin{cases} \omega & i = 0 \\ 0 & i \neq 0 \end{cases}$

- $\vee$ and $\wedge$ are sup and inf on $\omega + 1$, respectively.

Let us call the resulting system *Generalized Post logic*.

---

[*]By this formulation we mean that the truth value of atomic formulas is not known in the beginning, or, put another way, in *interpreted* reasoning we do not ask for tautologies, but for the truth value that a formula takes on under one or a set of given interpretations.

The proposition $D_i(p)$ essentially states that p has a truth value equal to or greater than i, whereas the proposition $\neg D_i(p) \cong D_i(p) \supset 0$ states that p has a truth value smaller than i. Note that the signs $\boxed{>i}$, $\boxed{<i}$ used in regular logics correspond exactly to the Post algebra operators D_{i+1} and $\neg D_i$.

The key point in making this logic applicable to resolution is the fact that each formula can be transformed into a CNF formula where only statements of that simple form can occur as literals:

Definition 8.30. (D-CNF) *A **L**-formula is said to be in **D-CNF** iff it is in CNF and all literals occurring in it are of the form $D_i(p)$ or $\neg D_j(q)$, where p, q are atomic.*

Theorem 8.31. (D-CNF (Rasiowa, 1973; Orłowska, 1978)) *For every closed **L**-formula ϕ, there exists a corresponding closed formula ψ in* **L** ^{sko} *which is in universal prenex D-CNF, such that $\phi \supset \psi$ and $\psi \supset \phi$ are first-order tautologies.*

Since a fully expanded semantic tableau can be seen as a DNF over its signed atomic formulas, it is possible to interpret an expanded tableau in a regular logic as a **D-DNF**, which is defined in analogy to D-CNF above. Hence, regular logics can be characterized as a class of logics that have 'uniform properties' with respect to Post algebras.

The finitely-valued version bears close resemblance to the logics considered by Morgan (see Section 8.1.1), which may also be expressed by Post algebras.

The main idea in the proof of the previous theorem is to make use of the equivalence

$$\phi \cong \bigvee_{i=1}^{\omega} (D_i(\phi) \wedge i)$$

and then to apply a vast number of transformation rules to yield the result.

This means that in general the D-CNF of a given formula involves infinitely many clauses. The proof of the Herbrand theorem in Orłowska (1978) shows that for any atom p, all but finitely many statements of the form $D_i(p)$ are redundant, and, hence, a finite resolution principle is sufficient.

If the D_i are interpreted as signs and a CNF algorithm is derived from inverse tableau rules for these signs like in Section 6.3.4 the transformation rules given in (Orłowska, 1978) could be greatly improved.

As a resolution rule one can take the usual binary resolution rule (Chang and Lee, 1973), with the following modified notion of a complementary pair.

Definition 8.32. (Complementary literals) *Let ϕ, ψ be two literals from a formula in D-CNF. They are **complementary literals** iff the following holds:*

1. *Exactly one of them has the form $D_i(p)$ and one has the form $\neg D_j(q)$; without loss of generality assume that $\phi = D_i(p)$ and $\psi = \neg D_j(q)$.*

2. *$j \leq i$.*

Note that if the D_i are interpreted as signed atoms the preceding definition boils down to the tableau closure condition in the presence of regular signs (cf. the remark after Definition 5.9). Adding the usual factoring rule (Chang and Lee, 1973) yields a complete resolution system.

Theorem 8.33. (Soundness, completeness (Orłowska, 1978))   *Let ϕ be an **L**-formula and ψ a universal prenex D-CNF of ϕ. Let E be the set of clauses corresponding to the matrix of ψ. Then ϕ is unsatisfiable iff $\Box$ can be obtained by a finite number of resolution steps from E.*

In (Orłowska, 1985) a resolution proof system for the finitely-valued version of this Post logic is given; the resolution rule turns out to be the same as in the $\omega + 1$ case. The differences lie in using only a subset of the operators in the finite case and in a slightly different algorithm for computing the normal form.

8.1.5   *Paraconsistent logics*

da Costa *et al.* (1990) and Lu *et al.* (1991) give a resolution procedure for certain *paraconsistent logics*. In paraconsistent logics it is possible to deal with inconsistency in a non-trivial way. The language and semantics is that of classical first-order logic for non-atomic formulas comprising $\lor$, $\land$, $\supset$, $\sim$, $\forall$, $\exists$ with their usual meaning. Atoms are *annotated predicates*, possibly prefixed by one or more special negation signs $\neg$. The annotation is a truth value of a complete lattice. The string $A : \mu$ where A is an atom and μ a truth value informally states that the truth value of A must be at least μ with respect to the lattice ordering. $\neg$ is understood as a lattice complement. Using the upset of μ in the lattice and interpreting the lattice elements as truth values we can express the signed literal $A : \mu$ in *paraconsistent logic* with the signed literal $\uparrow \mu\, A$ in *many-valued logic*. See Lu *et al.* (1993) for a formal treatment of this relationship between paraconsistent logic and many-valued logic.

In da Costa *et al.* (1990) a normal form theorem as well as a sound and complete resolution rule for this kind of logics is stated. The model theory is essentially two-valued and the truth value lattice is only used for evaluation of atoms.

In da Costa *et al.* (1990) and Lu *et al.* (1991) only upsets and complements of upsets are used as annotations of atoms. The existence of regular signs, however, suggests that upsets and downsets are at least as reasonable a choice for the set of annotations. Indeed, there seems to be a computational advantage in using upsets and downsets. This discussion is outside the scope of the present overview and will be conducted in a forthcoming paper (Hähnle *et al.*, 1993).

In Section 5.5 we sketch a way of dealing with paraconsistent logics in a tableau setting.

8.2   Other approaches

The first two approaches reviewed in the present section are loosely related (though they have been developed independently) through their relationship with semantic tableaux (cf. Section 3.1). The following three subsections are many-valued versions of completely different proof methods, while in the last subsection we mention two very general proof methods, which work for virtually any axiomatizable logic and in particular for many-valued logics.

8.2.1   *Decision diagrams*

8.2.1.1   *Binary decision diagrams*   This method is based on the *Binary Decision Diagram (BDD)* representation of classical propositional formulas. The main idea is to express any binary propositional function with a ternary `if-then-else` connective (see Table 8.1).

Table 8.1   *The truth table of the two-valued* `if-then-else`; *e is the first argument, p and q the second and third*

e/pq	00	01	10	11
0	0	1	0	1
1	0	0	1	1

Obviously, all unary and binary classical connectives can be expressed with the help of `if-then-else` and constants 0, 1, for example:

$$p \wedge q \cong \text{if } p \text{ then } q \text{ else } 0$$

$$p \vee q \cong \text{if } p \text{ then } 1 \text{ else } q$$

and so on. One can then show that a formula ϕ in reduced DAG representation that is composed exclusively of `if-then-else`, atomic formulas, and the primitives 0 and 1 has a strong normal form which is identical to 1 in the case where ϕ is a tautology.

BDDs in tree representation are sometimes also called Shannon graphs. The BDD technique belongs to the early lore of computer science and has been reinvented several times. It occurs in Shannon (1938) and Lee (1959) and became widely known via Akers (1978).

Recently, fast Boolean function manipulation packages for VLSI circuit design based on binary decision diagrams have become available with storage techniques that use the normal form properties of `if-then-else` expressions very efficiently (Billon and Madre, 1988; Brace *et al.*, 1990;

Bryant, 1992). Shannon graphs have been proposed as a theorem proving method for propositional logic in Ehrenfeucht and Orłowska (1967). This work has been extended to cover certain first-order logics (Orłowska, 1969a; Orłowska, 1969b) and finitely-valued propositional logic (Orłowska, 1967).* Recently, Posegga (1992a; 1992b) proposed an alternative extension to full first-order logic.

Instead of describing Orłowska's algorithm (1967) in detail, we present a short example for two-valued logic and point out what has to be changed to cover the many-valued case.

In Figure 8.1 we represent if e then r else s for some r, s graphically as a binary tree whose root is e; the right child (the edge labelled with 1) is the **then**-part and the left child the **else**-part. To show that $p \supset (q \supset p)$ is a tautology we make use of the equivalence

$$r \cong \text{if } r \text{ then } 1 \text{ else } 0$$

which holds for all r, and compute the Shannon graph in two steps (the leftmost tree in the second line). Next comes a simplification rule which exploits the fact that the value of a propositional variable is determined when it occurs the second time (in the example the lower p must be 1). This step is followed by two simplification steps, which are justified by the validity of

$$\text{if } r \text{ then } i \text{ else } i \cong i$$

Since all transformations preserve the truth value we have indeed proved that $p \supset (q \supset p)$ is a tautology.

8.2.1.2 *n-ary decision diagrams* Now we outline the necessary modifications that have to be made for finitely-valued propositional logics.

1. Instead of a `if-then-else` use a `case-of` statement that has as many cases as the logic has truth values, and express the many-valued connectives with the help of `case-of`. Our three-valued conjunction (see Table 2.1), for example, would read as the diagram depicted in Figure 8.2, where the `case-of` statement is represented as a n-ary tree and the P_i are arbitrary subtrees.
2. The rule for deletion of the second occurrence of identical variables is not changed.
3. The other simplification rule follows from the following equivalence:

---

*In all of latter papers the proof system is given in terms of Markov algorithms and the ideas are buried under a somewhat strange notation. Although Ehrenfeucht and Orłowska (1967) claim that they have implemented the propositional case, the given algorithm lacks several obvious efficiency improvements. The work in Ehrenfeucht and Orłowska (1967) and Orłowska (1969a; 1969b) has been carried out completely independently from other sources.

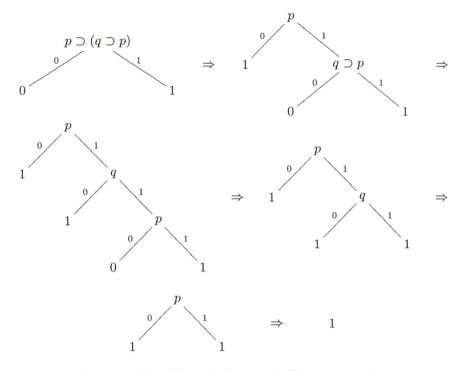

FIG. 8.1. Tautology checking with Shannon graphs.

```
case e of
     0    : i;
    1
   ───  : i;              ≅        i
   n-1
    ...
     1    : i
esac
```

The propositional language under consideration in Orłowska (1967) incorporates the connectives $\{\vee, \wedge, \supset, J_0, \ldots, J_1\}$ with similarity type $\langle 2, 2, 2, 1, \ldots, 1\rangle$. The semantics of $\vee, \wedge, J_0, \ldots, J_1$ is as in Definition 2.28, while $\supset$ is defined as

$$i \supset j = \left\{ \begin{array}{ll} 1 & i \leq j \\ 0 & i > j \end{array} \right.$$

The following theorems state soundness and completeness for these logics. Although these results are not stated explicitly in Orłowska (1967) they are easy consequences of other results in the same paper.

Theorem 8.34. (Soundness) *Let ϕ be a propositional formula and let $\mathcal{T}$ be the tree that corresponds to the formulas obtained after a number of*

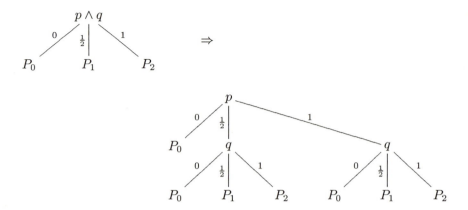

FIG. 8.2. Dealing with three-valued conjunction in n-ary decision diagrams.

transformation steps. Let I be the set of truth values occurring in the leaves of $\mathcal{T}$. Then $v(\phi) \in I$ for any valuation v.

As suggested by this theorem, the same technique can be used not only to check many-valued tautologies, but also to determine exactly the range of truth values that arbitrary valuations of a formula can have.

Theorem 8.35. (Completeness) *Let ϕ be a propositional formula and assume that $v(\phi) = i$ for every valuation v. Then ϕ can be reduced to i in a finite number of transformation steps.*

As can easily be seen, there is ample room for various improvements of the rules (in fact the rule for $\wedge$ given above is simpler than the one given in Orłowska (1967)), for example more sophisticated simplification rules could be designed. As in classical BDDs, one would use DAGs instead of trees in order to achieve a smaller representation. Also, there is no reason for separating expansion and simplification steps. On the contrary, we conjecture that the possibility of intermingling expansion and simplification rules lifts proof length complexity (Cook and Reckhow, 1974) in a similar way as does the admission of analytic cut in semantic tableaux (D'Agostino, 1990). See Section 6.3.2 for another improvement involving sets-as-signs.

8.2.1.3 *Multiple-valued deduction as a unification problem*   We mention another approach to many-valued deduction in this subsection which can be implemented on the basis of decision diagrams.

Recall that semantics of many-valued logics are provided by abstract algebras. Obviously, deduction problems in a propositional logic $\mathcal{L} = (\mathbf{L}, \mathbf{A})$, for example $\vdash \phi \cong \psi$, can be reformulated as a unification problem in the algebra $\mathbf{A}$, namely whether there is a most general unifier $\sigma : L_0 \to L$ such that $\phi\sigma =_{\mathbf{A}} \psi\sigma$. Provided that $\mathbf{L}$ is finitely-valued and functionally complete, an effective unification algorithm that computes the most general unifier of two formulas relative to $\mathbf{A}$ can be formulated (Nipkow, 1988). Nipkow's algorithms have been improved and applied to hardware verification (cf. Section 7.3.2) in Büttner *et al.* (1990). The implementation is done by first reducing the unification problem for a certain family $\mathbf{L}_n$ of functionally complete algebras to unification in Boolean algebras of size n. Boolean unification can then be implemented by n-ary decision diagrams. The universality of the approach stems from the fact that the $\mathbf{L}_n$ are functionally complete. On the other hand, functional completeness is necessary in any event for the existence of most general unifiers.

It turns out that the family of algebras used by Büttner *et al.* (1990) bears a very close resemblance to the family $\mathbf{L}_M^n$. As a consequence, the remarks on the restricted usability of $\mathbf{L}_M^n$ made at the end of Section 8.1.1 apply here also. Moreover, the approach is limited to propositional logic.

8.2.2 *Approaches based on tableaux and Gentzen calculi*

This is the approach that our own work starts from; therefore, we decided to integrate the presentation into our development at the proper place. We refer the reader to Section 3.3. For a brief introduction into the method of semantic tableaux, see Section 3.1. Introductions into Gentzen systems can be found in most logic textbooks, for example in Fitting (1990b).

On the other hand, we want to mention at least the, historically, most important papers which led to the development of tableau procedures for many-valued logics.

Probably the first Gentzen style axiomatization which was already completely general for arbitrary finitely-valued propositional logics (and even some first-order logics) was given by Schröter (1955). Rousseau (1967), independently, gave an almost identical Gentzen characterization of many-valued logics which also accounted for the general class of distribution quantifiers (cf. Section 3.3). Neither paper uses signs in order to stipulate truth values of formulas, but, rather, the syntactic position in generalized n-ary sequents. For example, in the sequent

$$\Gamma_0 | \Gamma_{\frac{1}{n-1}} | \cdots | \Gamma_1$$

a formula ϕ must take on truth value i iff $\phi \in \Gamma_i$. The notation used by Rousseau in some examples comes close to sets-as-signs, but his interest was purely proof theoretic and the notion was not developed.

Signs as a notational device for many-valued proof systems were introduced independently by Surma (1984) and Suchoń (1974) in the domain of semantic tableaux. Carnielli (1987) paralleled for semantic tableaux what Rousseau did for Gentzen systems, again completely independent from Rousseau's work. The Gentzen system in Baaz and Zach (1992) boils down to a special case of Rousseau (1967). The similarity between Gentzen and tableaux systems, as well as two possible and dual formulations of satisfiability in the many-valued case, are emphasized in Baaz *et al.* (1992).

The first paper involving a systematic treatment of truth value sets-assigns is Hähnle (1990b). Independent formulations for special cases were given in Doherty (1990a), Murray and Rosenthal (1991a), and Zach (1992).

8.2.3   *Path dissolution by Murray and Rosenthal*

Dissolution is a sound and complete inference rule for classical first-order logic that was introduced in 1986 by Murray and Rosenthal (1986; 1987); see Murray and Rosenthal (1990a; 1993a) for a detailed description. It operates on formulas in prenex negation normal form that are built up from *n*-ary conjunctions and disjunctions. It bears some similarity to Bibel's connection method (Bibel, 1987) in that the focus is on maximal conjunctive paths through the formula tree. It can also be seen as a refinement of a proof method by Prawitz (1970).

In order to keep things simple, we constrain our exposition to the propositional case. Since dissolution is not so widely known as yet, we begin with a small example from classical logic before we turn to many-valued logics. Consider a formula built up from $\land, \lor$, and $\neg$ in NNF, which is represented as a graph, where nodes are formulas and edges are either of conjunctive or disjunctive type.

In the left part of Figure 8.3 we have drawn the graph for $\phi = D \land (A \lor B) \land (\overline{A} \lor C)$ (where $\neg$ is denoted by $^{-}$). Conjunctive connections are drawn vertically, while disjunctive connections are drawn horizontally.

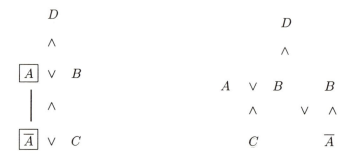

FIG. 8.3. Dissolution for classical propositional logic.

Now consider maximal sets of conjunctively connected literals, so-called *c-paths** in ϕ. For the present example, these are

---

*We will not define the graph-based notions that are used in the following formally. The intuitive reading should be clear and we refer the reader to Murray and Rosenthal (1990a) for the exact definitions.

$$\{\{D, A, \overline{A}\}, \{D, A, C\}, \{D, B, \overline{A}\}, \{D, B, C\}\}.$$

A pair of complementary literals lying on the same c-path is called a *link*. The dissolution rule always operates on a link in focus.* The central idea behind dissolution is to restructure a formula in such a way that exactly the c-paths containing the link in focus are removed. The right-hand part of Figure 8.3 shows ϕ after dissolving on the link $(A, \overline{A})$, which is highlighted on the left-hand side. One observes that the set of c-paths is now

$$\{\{D, A, C\}, \{D, B, \overline{A}\}, \{D, B, C\}\},$$

where the one path containing $(A, \overline{A})$ has been removed. The completeness of the method in the propositional case now follows from the fact that after a finite number of steps there can be no more c-paths left that contain any links. Since in an unsatisfiable formula each c-path must contain at least one link, the empty graph must be produced after a finite number of dissolution steps if and only if the starting formula was unsatisfiable.

Since NNF representation of formulas plays a crucial role in dissolution, the applicability to many-valued logics in the present form is restricted to logics that support the computation of NNFs. Recently, Murray and Rosenthal (1991c; 1993b) presented a technique eliminating the dependence on specific operators by translating any finitely-valued logic into a classical semantic And/Or-graph with general signs. The method is closely related to the techniques developed in Chapters 4 and 5 and we discuss this relationship and some possible improvements in Section 6.3.3.

Definition 8.36. (UNF logic) *A propositional logic* $\mathcal{L}_{\mathrm{UNF}}^n$ *is called a* **unary normal form (UNF) logic** *iff the following conditions are satisfied:*

- *The language of* $\mathcal{L}_{\mathrm{UNF}}^n$ *is* $(L_{UNF}^n, \vee, \wedge, o_1, \ldots, o_k)$ *with similarity type* $\langle 2, 2, 1, \ldots, 1 \rangle$.
- $\vee$, $\wedge$ *are defined as usual.*
- $\{o_1, \ldots, o_k\}$ *are defined in such a way that every* $\mathbf{L}_{\mathrm{UNF}}^n$*-formula is equivalent[†] to a formula in unary normal form, i.e. one which is composed solely of literals,* $\wedge$ *and* $\vee$.

All logics mentioned in the preceding sections of the present chapter (with the exception of Section 8.1.3) in connection with resolution systems are UNF logics.

---

*In practice, that is. In general, a multiple-link dissolution rule can be defined. The multiple-link rule also represents an alternative approach to many-valued dissolution. It turns out, however, that implementation and control of multiple-link rules is not feasible in practice and hence will not be considered here.

[†]It suffices to stipulate refutational equivalence here.

The main obstacle in developing a many-valued dissolution rule is the fact that in the many-valued case links are not necessarily binary. For example, the set of literals* $\{\sim A, \sim \neg A\}$ does not constitute a link (or a contradiction). One solution would be to design a dissolution rule that can handle n-ary links, an approach that results in many technical difficulties. Thus in Murray and Rosenthal (1991b) the dissolution rule is extended to handle incomplete or *partial links*.

The basic idea (inspired by Hähnle (1990b) and Doherty (1990a), see also Chapter 4) is to substitute for each literal $o_1 \cdots o_m p$ in an NNF formula a *signed* literal of the form Sp, where S is the set of truth values for which $v(o_1 \cdots o_m p) \in D$ iff $v(p) \in S$. For example, $\sim \neg p$ would be substituted by $\{\frac{1}{2}, 1\}$. Therefore, in the dissolution step the c-path containing a partial link $\{S_1 A, S_2 A\}$ is not deleted, but is replaced by a path that contains $\{(S_1 \cap S_2) A\}$ in place of the two literals.

We illustrate this by giving a derivation based on many-valued dissolution of our earlier example.

Example 8.37. *We prove that* $\sim (\neg p \supset (\sim p \wedge \neg p))$ *is not satisfiable. First substitute* $\supset$ *by* $\sim$ *and* $\vee$:

$$\sim (\neg p \supset (\sim p \wedge \neg p))$$
$$\equiv \quad \sim (\sim \neg p \vee (\sim p \wedge \neg p))$$

Next, we compute UNF:

$$\equiv \quad \sim\sim \neg p \wedge (\sim\sim p \vee \sim \neg p)$$

Substituting truth value sets for unary operators yields

$$\equiv \quad \{0\}\, p \wedge (\{1\}\, p \vee \{\tfrac{1}{2}, 1\}\, p)$$

If we represent this graphically, we obtain the left-hand part of Figure 8.4.

Dissolving on the partial link (which is also a full link) $\{\{0\}\, p, \{1\}\, p\}$ *results in the graph displayed at the right-hand side, from where the empty graph can be derived in one more step.*

We should add that in the many-valued version of dissolution the number of c-paths does not necessarily decrease from one dissolution step to

---

*The operators $\neg$, $\sim$ are strong and weak negation as defined in Section 2.3.1.

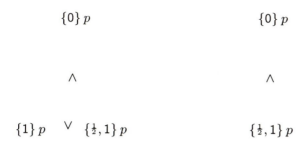

$$\{0\}\,p \qquad\qquad\qquad \{0\}\,p$$

$$\wedge \qquad\qquad\qquad\qquad \wedge$$

$$\{1\}\,p \quad \vee \quad \{\tfrac{1}{2},1\}\,p \qquad\qquad\qquad \{\tfrac{1}{2},1\}\,p$$

FIG. 8.4. Example of many-valued dissolution.

the next, however, the number of partial links which belong to a c-path decreases.

Murray and Rosenthal (1990b) prove that a restricted version of the dissolution rule can be used to improve the complexity of proof length (Cook and Reckhow, 1974) of propositional formulas within semantic tableaux systems. In Murray and Rosenthal (1991a) it is proved that under certain restrictions the same is possible for many-valued logics. See Chapter 6 for a further discussion of this topic.

Dissolution seems to be a promising technique for theorem proving in many-valued logics. An extension to many-valued first-order logic along the lines given in Murray and Rosenthal (1990a) should not be difficult. The restriction to UNF logics is severe, but in Murray and Rosenthal (1991c) and Section 6.3.3 of this work a method for overcoming this severity is proposed.

8.2.4   *Beavers' approach to Łukasiewicz logic*

Consider infinitely-valued Łukasiewicz logic $\mathcal{L}_\omega$.

The close connection between certain function spaces and many-valued logics is exploited in the proof procedure for $\mathcal{L}_\omega$ by Beavers (1991). By McNaughton's theorem (McNaughton, 1951) there is a one-to-one correspondence between n-ary logical functions of $\mathcal{L}_\omega$ and continuous, piecewise linear functions with integral coefficients from the n-dimensional unit cube into $[0,1]$. These functions are called *McNaughton functions*.

If we denote the McNaughton function of a k-variable $\mathbf{L}_\omega$-formula $\phi(p_1,\dots,p_k)$ by $f_\phi(x_1,\dots,x_k)$, it is not difficult to prove that the domain of f_ϕ can be divided into finitely many subdomains, say $[0,1]^k = D_1\dot\cup\cdots\dot\cup D_j$, such that for all i the following hold:

- There is a linear function, say $\pi_i(x_1,\dots,x_k) = a_i + a_{1i}x_1 + \cdots + a_{ki}x_k$, with $\pi_i\!\restriction_{D_i} = f_\phi\!\restriction_{D_i}$.
- The boundary of D_i is defined by $m_i \leq j$ many linear inequalities in

k variables, say,

$$b_{11}x_1 + \cdots + b_{k1}x_k \;\leq\; b_1$$

$$\cdots$$

$$b_{1m_i}x_1 + \cdots + b_{km_i}x_k \;\leq\; b_{m_i}$$

- All a_i, b_j, a_{kl}, b_{mn} given above are integers.

Proving validity of a $\mathbf{L}_\omega$-formula ϕ is equivalent to checking whether f_ϕ is identical to 1 on $[0,1]^k$. From the considerations above (and continuity of f_ϕ) it follows that it is sufficient to show that for all i, π_i has 1 as its minimum value on D_i, which is in turn equivalent to solving j linear programming (LP) problems, each in k variables.

It is also fairly obvious that restricting the LP problem to the set $\{0, \frac{1}{n-1}, \ldots, \frac{n-2}{n-1}, 1\}^k$ yields a proof procedure for the Łukasiewicz logics with truth value set $\{0, \frac{1}{n-1}, \ldots, \frac{n-2}{n-1}, 1\}$ (note, however, that the LP problems then become IP problems and are much more difficult to solve).

Although LP is solvable in polynomial time, the algorithm itself is not necessarily polynomial. The reason is that the upper bound for the number of different subdomains is, in general, exponential in the number of binary connectives occurring in the formula, which is not surprising, since the satisfiability problem for $\mathcal{L}_\omega$ is known to be NP-hard (Mundici, 1987a) (cf. also Section 6.2). In practice, however, many subdomains turn out to be degenerate, i.e. they have a dimension of less than k, in which case they need not be considered.

The proof method can be improved in various details and may be extended to include other binary connectives definable in $\mathcal{L}_\omega$, such as $\vee$. It cannot easily be adopted to other logics, however, since it depends on certain characteristics of the representation functions of formulas. Beavers (1991) reports that he has written a FORTRAN77 program based on the technique, however; he does not provide test data.

8.2.5 *Mellouli's three-valued extension of Plaisted's modified problem reduction format*

Plaisted (1988) introduced the *modified problem reduction format (MPR)*, a first-order proof procedure which can be classified into the various attempts to extend Prolog-style Horn clause[*] logic programming to full first-order logic.

We will use some common logic programming terminology in this section. Readers who are unfamiliar with these notions may consult, for example, Lloyd (1987) for background.

---

[*]Horn clauses are clauses that contain at most one positive literal.

MPR has been simplified and extended to the logic $\mathcal{L}_3$ from Section 8.1.2 in Mellouli (1990). Mellouli also provided a Prolog-based MPR implementation of both classical first-order logic and $\mathcal{L}_3$.

MPR assumes that the input is given in universal prenex CNF. First we sketch the classical case. Construction of proof trees proceeds quite similarly as in Prolog: definite Horn clauses* are handled in exactly the same way as in Prolog. Clauses with no positive literals are rewritten into the form[†]

$$\bot :- l_1, \ldots, l_n.$$

where $\bot$ is a reserved positive literal denoting falsity. All other (non-Horn) clauses have the form

$$l :- \neg l_1, \ldots, \neg l_i, l_{i+1}, \ldots, l_n.$$

for some $i \geq 1$. A proof tree starts with root $\bot$ and is expanded by treating clauses as Prolog-like rules. One possibility for closing a branch is when a fact which is already on the branch is in the database. The most interesting detail, of course, is the handling of negative literals.

To this end three additional rules are provided. We assume that each branch of the proof tree has associated with it a list Γ of assumptions. The assumptions are attached to labelled nodes in the tree. Nodes are labelled with literals or have the empty label.

Assumption addition If the current branch ends with a negative literal $\neg p$ and $\{p, \neg p\}$ are not in the current list of assumptions, then one may attach $\neg p$ as an assumption below all other assumptions on the current branch and close the current branch.

This rule corresponds to the usual negation as failure rule, which is unsound. Soundness is achieved by the following rule.

Case analysis To any negative assumption $\neg p$ attached to a literal node a case analysis may be applied by creating a new child of this node which is unlabelled and has the assumption p attached to it. In a complete proof each assumption must have been analysed. This requirement makes assumption addition sound.

Assumption application Finally, assumptions that have been derived may be cancelled, that is, when a node literal unifies with an assumption on the current branch, the unifier is applied to the proof and the branch is closed.

---

*Definite Horn clauses are clauses that contain exactly one positive literal.

[†]In the following, $p :- q$ may be seen as an abbreviation for $p \vee \sim q$.

Before we turn to an example let us note the changes that have to be made for $\mathcal{L}_3$. The normal form that is being used is S-CNF from Section 8.1.2. Rules and proof trees can be constructed analogous to the classical case, when literals of the form $p, \sim \neg p$ are treated as positive and literals of the form $\neg p, \sim p$ as negative. For the sake of clarity we repeat the construction rules including the modifications made for $\mathcal{L}_3$.

Assumption addition If the current branch ends with a negative literal $\neg p$ (resp., $\sim p$) and $\{p, \neg p, \sim \neg p\}$ (resp., $\{p, \sim p\}$) are not in the current list of assumptions, then attach $\neg p$ (resp., $\sim p$) as an assumption below all other assumptions on the current branch and close the current branch.

Case analysis To any negative assumption $\neg p$ (resp., $\sim p$) attached to a literal node a case analysis may be applied by creating a new child of this node which is unlabelled and has the assumption $\sim \neg p$ (resp., p) attached to it. In a complete proof each assumption must have been analysed.

Assumption application Unchanged.

For completeness a kind of consistency or weakening rule has to be added:

$$\sim (p \wedge \neg p) \cong \sim \neg p :- p$$

We illustrate the method by providing one more refutation of our example $\neg p \supset (\sim p \wedge \neg p)$. Remember that the transformation to S-CNF yielded:
(1) $\neg p$,
(2) $p \vee \sim \neg p$.
After transformation to rule notation we have
(1′) $\bot :- \sim \neg p$,
(2′) $p :- \neg p$.
We obtain the following proof tree (assumptions are written in parentheses to the left of the node to which they belong, node labels are on the right):

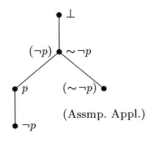

(Assmp. Add.)

Mellouli observed that removal of the consistency rule results in a proof system for a somewhat weaker four-valued logic. Its adoption to other many-valued logics, on the other hand, is far from obvious, if at all possible. Thus the proof method is basically a special purpose implementation for $\mathcal{L}_3$. Its merits lie in the closeness to standard Prolog: Mellouli's implementation based on Quintus Prolog has similar characteristics to other Prolog-based theorem provers such as Stickel's PTTP (Stickel, 1988), Manthey and Bry's SATCHMO (1988), and, of course, Plaisted's original MPR (Plaisted, 1988).

8.2.6   General frameworks

For the sake of completeness we mention two general purpose approaches to mechanical theorem proving, in which many-valued logics merely form a special case.

8.2.6.1   AUTOLOGIC

Morgan (1985) presented the theorem prover AUTOLOGIC that works in principle for arbitrary propositional logics whenever a Hilbert-style axiomatization is available, that is, a finite number of axiom schemata and inference rules (which take a finite number of formulas as premisses and have a single formula as conclusion) which characterize the tautologies of the logic. Starting from the theorem to be proved, the system works by generating a backward proof tree until each leaf matches with an axiom. It is, to best of my knowledge, the only serious attempt to construct a theorem prover based on Hilbert-style proof systems. Morgan claims that the system performs surprisingly well in comparison with the resolution-based Argonne Theorem Prover ITP (Wos *et al.*, 1984), at least for small knowledge bases, although he admits to haveing tested it on only a very small problem set. Morgan starts with a simple backward chaining algorithm which is subsequently improved by various optimizations. For our purposes the approach gains its relevance from the fact that for most propositional many-valued logics Hilbert-style axiomatizations are available (Rescher, 1969). On the other hand, due to indeterminism the search space is enormous and without specifying suitable lemmata, anti-lemmata, and derived rules the automated finding of more complex problems seems improbable, especially since Hilbert-style proofs are in practice long and tedious in comparison to their counterparts in other proof systems.* Also, dis-proofs or counterexamples cannot usually be detected using this approach, added to which the extension to first-order logic is not easy to achieve.

8.2.6.2   Labelled deductive systems

Gabbay (1991) developed a proof theoretic framework, called *Labelled Deductive Systems (LDS)*, which covers a very wide range of non-standard logics including, for example, rele-

---

*This is not true in theory: Hilbert systems have a smaller proof length complexity than the inference systems generally used in automated theorem proving.

vant, many-valued or non-monotonic logics, both in the propositional and first-order case. It is well beyond the scope of this brief overview to even give a survey of LDS however we highlight the basic features.

The central idea is to make semantical or meta level information explicit in a formal derivation instead of hiding it in the pattern of axioms and rules. The technical representation is done via signs or *labels* that are attached to each formula. This makes it possible to fix a set of logical rules (that manipulate formulas) for all kinds of logics (that have, of course, to share the same formula language) and incorporate the semantical information in a systematic way into the manipulation of labels. The language of labels can be freely chosen and a label intuitively represents the status of the formula that it is attached to in the current derivation, for example it may contain information on resources used to derive it, or on its reliability, or whatever. From a proof theoretical point of view LDS look very much like natural deduction systems, but with the extra feature of a labelling discipline that specifies the initial labelling and the propagation of labels. The considerable flexibility of LDS comes from the fact that both information on the semantic models and proof theoretic subtleties of natural deduction systems can be combined in the handling of labels.

Automation has not been an issue so far in the development of LDS, so in order to provide some arguments we mention the LDS framework here. First, we expect that in the future LDS will have a considerable influence on some of the research being undertaken in the area that can be described as *computational logic for AI applications* and, therefore, implementation *will* become an issue sooner or later. Second, even if LDS are not directly suited for implementation they are a tool for gaining insight into the computational properties of non-classical logics and this insight can then be used to adapt a logical framework which is more suitable for automation than LDS. This philosophical standpoint is similar to our use of the tableau framework in the present work. Of course, the gap between theory and implementation is much broader with LDS than with tableaux, but then LDS are considerably more general. It is conceivable that tableaux are used as a kind of target system for the (prototypic) implementation of specific LDS. First results in this direction are reported in D'Agostino and Gabbay (1993).

8.3   Discussion

The purpose of this chapter is to give an overview of the work done in many-valued theorem proving alongside the efforts presented earlier in this book. It is fairly obvious that most of the approaches reviewed are of historical interest by now. Still, there may be something to be learned. One conclusion to be drawn is that several techniques, for example the many-valued resolution rule or the DNF representation of many-valued truth conditions with signs and other syntactic means were reinvented time and again. Several completely independent completeness proofs of many-valued resolution which differ only in minor details are still around. One aim of this historical account is to decrease the degree of redundancy that permeates the research undertaken in this field. Another achievement of such an historical account could lie in a possible comparison of the various approaches in order to rule out the ones which are not likely to be successful. Let us look at the catalogue of evaluation criteria for many-valued inference systems again.

Wide applicability

This requirement excludes all special purpose systems.

This leaves us with decision diagrams, path dissolution, non-clausal resolution; resolution methods based on signed clauses also are a candidate if a transformation algorithm to CNF as mentioned in Section 6.3.4 is used. It is true that some of the logics considered in the other approaches are functionally complete in the propositional case and can thus, in principle, simulate all other logics, but as we have already pointed out in Section 8.1.1, functional completeness alone does not suffice in practice. One should add that first-order logics are difficult to cover in AUTOLOGIC.

Flexibility

This point is not applicable to specialized approaches.

Easy Adaptability

Adapting LDS to a new logic is in general an ingenious process, while in AUTOLOGIC it depends on the availability of a Hilbert-style axiomatization.

Performance

We refer to the discussion of this point in Section 6.4.1.

We may add that the number of possible resolvents that form the search space may just be too large within the framework of Stachniak. The MPR approach performs excellently for the special case for which it has been designed. AUTOLOGIC is probably limited to relatively simple problems.

Closeness to classical versions

For Stachniak's resolution and for decision diagrams, significant new implementation problems have to be faced. For LDS there exists no automated theorem proving paradigm so far; this also applies to Beavers' approach, whose implementation seems to be rather *ad hoc*.

9

CONCLUSION

'If we could only find some absolutely impossible...'
'Unwritable...'
'Unfinishable...'
'Unimaginable...'
'Endlessly revisable...'
'Text without words...'
— John Fowles, *Mantissa*

In the preceding chapters we presented a new approach to many-valued theorem proving starting out from signed analytic tableaux. The main technical device was a systematic extension of the sign language. We showed that admitting subsets of the set of truth values of a logic as signs is sufficient to build compact and efficient tableau systems for a large class of many-valued logics and we coined the term *sets-as-signs approach* for that technique (Chapter 4). Its extendability to many-valued first-order logics was demonstrated (Section 5.4).

The subclass of regular logics was defined (Section 5.2) and it was conjectured that these logics correspond precisely to those that have a uniform notation style system when the sets-as-signs approach is used (Section 5.5). We provided evidence that our sets-as-signs approach can easily be extended and used for logics whose connectives have a semantics based on a partial order on the truth values, thus yielding even more simple and efficient tableaux systems for such logics (Section 5.5).

On the other hand, we proved that the usefulness of the sets-as-signs approach is restricted neither to analytic tableaux nor to many-valued logics. Most semantically oriented proof procedures can be extended to cover many-valued logics using the sets-as-signs technique (Sections 6.1, 6.3.3). In Section 6.2 we further generalized our ideas and introduced tableau rules with variables in the signs and annotated integer and linear constraints. As a result, a new translation from logic into mixed linear and integer programming is achieved. This enables one to undertake automated theorem proving in infinitely-valued propositional logics and provides for the first time an efficient method of handling Łukasiewicz logics. A corollary of the mixed integer translation is the availability of easy NP-containment proofs for a wide variety of finitely and infinitely-valued logics.

Previous approaches to many-valued theorem proving were exhaustively reviewed (Chapter 8) and, as a byproduct, the most complete bibliography on the topic to date has been accumulated.

In Sections 7.2.1, 7.2.2, 7.2.4, 7.3.1, 7.3.2 we collected examples illustrating that a flexible and general theorem prover is indeed desirable.

The next steps now are (i) to construct a competitive many-valued theorem proving system along the lines sketched at the end of Section 6.4.3, and (ii) to use such a system in order to move from academic examples to real-world applications.

REFERENCES

Ackermann, R. (1967). *Introduction to Many-Valued Logics*. Routledge & Kegan, London.

Akers, S. B. (1978). Binary decision diagrams. *IEEE Transactions on Computers*, **27**(6), 509–516.

Anderson, A. R., and Belnap Jr, N. D. (1975). *Entailment: The Logic of Relevance and Necessity*. Vol. 1. Princeton University Press.

Andrews, P. B. (1981). Theorem proving through general matings. *JACM*, **28**, 193–214.

Andrews, P. B. (1986). *An Introduction to Mathematical Logic and Type Theory*. Academic Press.

Baaz, M., and Fermüller, C. G. (1992). Resolution for many-valued logics. *Pages 107–118 of:* Voronkov, A. (ed.), *Proc. Logic Programming and Automated Reasoning LPAR'92*. Springer, LNAI 624.

Baaz, M., and Zach, R. (1992). Note on calculi for a three-valued logic for logic programming. *Bulletin of the EATCS*, **48**(oct), 157–164.

Baaz, M., Fermüller, C. G., and Zach, R. (1992). Dual systems of sequents and tableaux for many-valued logics. *In: Proc. 2nd Worshop on Theorem Proving with Tableaux and Related Methods, Marseille*. Tech. Report, MPI Saarbrücken.

Balas, E., Ceria, S., and Cornuéjols, G. (1993). Solving mixed 0–1 programs by a lift-and-project method. *Pages 232–242 of: Proc. of the ACM-SIAM Symposium on Discrete Algorithms, Austin, Texas*.

Balbes, R., and Dwinger, P. (1974). *Distributive Lattices*. University of Missouri Press, Columbia.

Beavers, M. G. (1991). Automated Theorem Proving for Łukasiewicz Logics. Manuscript of talk given at 1991 Meeting of Society for Exact Philosophy, Victoria, Canada.

Beckert, B. (1992). Computational Complexity: Semantic Tableaux with Lemma Generation vs. D'Agostino's KE-System. Technical Note 10 TCG Project.

Beckert, B., Hähnle, R., and Schmitt, P. H. (1993). The *even more* liberalized δ-rule in free variable semantic tableaux. *In: Proceedings of the Kurt Gödel Conference, Brno, Czech Republic (to appear)*. Springer LNCS.

Belnap Jr., N. D. (1977). A useful four-valued logic. *Pages 8–37 of:* Dunn, J. M., and Epstein, G. (eds.), *Modern uses of multiple-valued logic*. Reidel, Dordrecht.

Beth, E. W. (1986). Semantic entailment and formal derivability. *Pages*

262–266 of: Berka, K., and Kreiser, L. (eds.), *Logik-Texte. Kommentierte Auswahl zur Geschichte der modernen Logik.* Berlin: Akademie-Verlag.

Bibel, W. (1980). *The Complete Theoretical Basis for the Systematic Proof Method.* Technical report. Forschungsgruppe Intellektik, Universität München.

Bibel, W. (1987). *Automated Theorem Proving.* Second revised edn. Vieweg, Braunschweig.

Billon, J., and Madre, J. (1988). Original concepts of PROAM, an industrial tool for efficient formal verification of combinatorial circuits. *Pages 487–501 of: Proceedings WG.10.2 Working Conference on the Fusion of Hardware Design and Verification.*

Bittencourt, G. (1989). A four-valued semantics for a n-valued terminological language. *In: Proc. Second International Symposium on Artificial Intelligence, Monterrey, Mexico.*

Blau, U. (1978). *Die dreiwertige Logik der Sprache: ihre Syntax, Semantik und Anwendung in der Sprachanalyse.* De Gruyter, Berlin.

Blikle, A. (1991). Three-valued predicates for software specification and validation. *Fundamenta Informaticae,* **XIV**, 387–410.

Bolc, L., and Borowik, P. (1992). *Many-Valued Logics. 1: Theoretical Foundations.* Springer Verlag.

Boy de la Tour, T. (1990). Minimizing the number of clauses by renaming. *Pages 558–572 of:* Stickel, M. E. (ed.), *Proc. 10th International Conference on Automated Deduction, Kaiserslautern.* Springer, LNCS 449.

Brace, K. S., Rudell, R. L., and Bryant, R. E. (1990). Efficient implementation of a BDD package. *Pages 40–45 of: Proc. 27th ACM/IEEE Design Automation Conference.* IEEE Press.

Bryant, R. E., and Seger, C.-J. H. (1991). Formal verification of digital circuits using symbolic ternary system models. *Pages 33–43 of:* Clarke, E. M., and Kurshan, R. P. (eds.), *Computer-Aided Verification: Proc. of the 2nd International Conference CAV'90.* Berlin, Heidelberg: Springer.

Bryant, R. Y. (1984). A switch-level model and simulator for MOS digital systems. *IEEE Transactions on Computers,* **C-33**, 160–169.

Bryant, R. Y. (1986). Graph–based algorithms for Boolean function manipulation. *IEEE Transactions on Computers,* **C-35**, 677–691.

Bryant, R. Y. (1987). A survey of Switch-Level algorithms. *IEEE Design and Test,* August, 26–40.

Bryant, R. Y. (1992). *Symbolic Boolean Manipulation with Ordered Binary Decision Diagrams.* Technical report. Carnegie Mellon University. School of Computer Science.

Brzozowski, J. A., and Seger, C.-J. (1991). Advances in asynchronous circuit theory. part II: Bounded inertial delay models, MOS circuits, design techniques. *Bulletin of the EATCS*, **43**, 199 –263.

Buro, M., and Büning, H. K. (1992). *Report on a SAT Competition.* Reihe Informatik 110. FB 17—Mathematik/Informatik, Universität Paderborn.

Büttner, W., Estenfeld, K., Schmid, R., Schneider, H.-A., and Tidén, E. (1990). Symbolic constraint handling through unification in finite algebras. *Applicable Algebra in Engineering, Communication and Computing*, **1**, 97–118.

Caferra, R., and Zabel, N. (1990). An application of many-valued logic to decide propositional S_5 formulae: a strategy designed for a parameterized tableaux-based theorem prover. *Pages 23-32 of: Proc. AIMSA'90, Artificial Intelligence—Methodology Systems Application.*

Carnielli, W. A. (1987). Systematization of finite many-valued logics through the method of tableaux. *Journal of Symbolic Logic*, **52**(2), 473–493.

Carnielli, W. A. (1991). On sequents and tableaux for many-valued logics. *Journal of Non-Classical Logic*, **8**(1), 59–76.

Chang, C.-L., and Lee, R. C.-T. (1973). *Symbolic Logic and Mechanical Theorem Proving.* Academic Press, London.

Cho, K., and Bryant, R. E. (1989). Test pattern generation for sequential MOS circuits by symbolic fault simulation. *Pages 418-423 of: 26th ACM/IEEE Design Automation Conference.*

Cohn, P. M. (1981). *Universal Algebra.* second edn. Reidel, Dordrecht.

Cook, S., and Reckhow, R. (1974). On the lengths of proofs in the propositional calculus. *Pages 135-148 of: Proceedings 6th STOC.*

da Costa, N. C. A., Henschen, L. J., Lu, J., and Subrahmanian, V. S. (1990). Automatic theorem proving in paraconsistent logics: Theory and implementation. *Pages 72-86 of:* Stickel, M. E. (ed.), *Proc. CADE-10.* Springer LNCS 449.

D'Agostino, M. (1990). Investigations into the Complexity of some Propositional Calculi. Ph.D. thesis, Oxford University Computing Laboratory, Programming Research Group. Also Technical Monograph PRG–88, Oxford University Computing Laboratory.

D'Agostino, M. (1992). Are tableaux an improvement on truth tables? Cut-free proofs and bivalence. *Journal of Logic, Language and Information*, **1**, 235–252.

D'Agostino, M., and Gabbay, D. M. (1993). Labelled refutation sustems: A case study. *In: Proc. Second Workshop on Theorem Proving with Tableaux and Related Methods, Marseille.* Technical Report, MPI Saarbrücken.

Davey, M., and Priestly, R. (1990). *Introduction to Lattices and Order.* Cambridge University Press.

de Bessonet, C. G. (1991). *A Many-Valued Approach to Deduction and Reasoning for Artificial Intelligence.* Kluwer Academic Publishers.

Di Zenzo, S. (1988). A many-valued logic for approximate reasoning. *IBM Journal of Research and Development*, **32**(4), 552–565.

Doherty, P. (1990a). Preliminary report: NM3—a three-valued non-monotonic formalism. *Pages 498–505 of:* Ras, Z., Zemankova, M., and Emrich, M. (eds.), *Proc. of 5th Int. Symposium on Methodologies for Intelligent Systems, Knoxville, TN.* North-Holland.

Doherty, P. (1990b). A Three-Valued Approach to Non-Monotonic Reasoning. Master's thesis, School of Engineering at Linköping University, Sweden.

Doherty, P. (1991). A constraint-based approach to proof procedures for multi-valued logics. *In: First World Conference on the Fundamentals of Artificial Intelligence WOCFAI-91, Paris.*

Doherty, P., and Lukaszewicz, W. (1991). *NML3—A Non-Monotonic Logic with Explicit Defaults.* Research Report, RKLLAB LiTH-IDA-R-91-13. Department of Computer and Information Science, Linköping University.

Dueck, G. W. (1988). Algorithms for the Minimization of Binary and Multiple-Valued Logic Functions. Ph.D. thesis, University of Manitoba, Winnipeg.

Dunn, J. M., and Epstein, G. (eds.). (1977). *Modern Uses of Multiple-Valued Logic.* Reidel, Dordrecht. Invited Papers of 5th ISMVL Symposium 1975 with Bibliography by R. G. Wolf.

Eder, E. (1991). Consolation and its relation with resolution. *In: Proc. IJCAI, International Joint Conference on Artificial Intelligence.*

Ehrenfeucht, A., and Orłowska, E. (1967). Mechanical proof procedure for propositional calculus. *Bull. de L'Acad. Pol. des Sci., Série des sci. math., astr. et phys.*, **XV**(1), 25 –30.

Epstein, G. (1960). The lattice theory of Post algebras. *Transactions of the American Mathematical Society*, **95**(2), 300–317.

Fenstad, J. E., Halvorsen, P.-K., Langholm, T., and van Benthem, J. F. A. K. (1985). *Equations, Schemata and Situations: A Framework for Linguistic Semantics.* Technical Report CSLI-85-29. Center for the Studies of Language and Information Stanford.

Fenstad, J. E., Langholm, T., and Vestre, E. (1988). Representations and interpretations. *In: Proceedings Workshop on Computational Linguistics and Formal Semantics Lugano.*

Filkorn, T., Schmid, R., Tidén, E., and Warkentin, P. (1991). Experiences from a large industrial circuit design application. *Pages 581–595 of:* Saraswat, V., and Ueda, K. (eds.), *Logic Programming: Proc. of the 1991 International Symposium.* Cambridge, MA: MIT Press.

Fitting, M. C. (1985). A Kripke-Kleene semantics for logic programming. *Journal of Logic Programming*, **4**, 295–312.

Fitting, M. C. (1986). Partial models and logic programming. *Theoretical Computer Science*, **48**, 229–255.

Fitting, M. C. (1988). Programming on a topological bilattice. *Fundamenta Informaticae*, **XI**, 209–218.

Fitting, M. C. (1989). Bilattices and the semantics of logic programming. *Journal of Logic Programming*, **11**(2), 91–116.

Fitting, M. C. (1990a). Bilattices in logic programming. *Pages 238–247 of: 20th International Symposium on Multiple-Valued Logic, Charlotte.*

Fitting, M. C. (1990b). *First-Order Logic and Automated Theorem Proving.* Springer, New York.

Fitting, M. C., and Ben-Jacob, M. (1990). Stratified, weak stratified and three-valued semantics. *Fundamenta Informaticae*, **XIII**, 19–33.

Gabbay, D. M. (1991). *LDS—Labelled Deductive Systems.* Oxford University Press, to appear. 6th Draft—February 1991.

Gerberding, S. (1991). Monomorphe Axiomatisierung von Intervallarithmetiken mit mehrwertigen Logiken. Master's thesis, University of Karlsruhe, Department of Computer Science.

Ginsberg, M. L. (1986a). *Multi-Valued Logics.* Technical Report TR-86-29. Department of Computer Science, Stanford University.

Ginsberg, M. L. (1986b). Multi-valued logics. *Pages 243–247 of: Proceedings of AAAI-86, 5th National Conference on Artificial Intelligence.* Morgan Kaufmann Publishers.

Gottwald, S. (1989). *Mehrwertige Logik. Eine Einführung in Theorie und Anwendungen.* Akademie-Verlag Berlin.

Grundy, M. (1990). Theorem Prover Generation Using Refutation Procedures. Ph.D. thesis, Department of Computer Science, University of Sidney.

Hähnle, R. (1990a). *Spezifikation eines Theorembeweisers für dreiwertige First-Order Logik.* IWBS Report 136. Wissenschaftliches Zentrum, IWBS, IBM Deutschland.

Hähnle, R. (1990b). Towards an efficient tableau proof procedure for multiple-valued logics. *Pages 248–260 of: Proceedings Workshop on Computer Science Logic, Heidelberg.* Springer, LNCS 533.

Hähnle, R. (1991). Uniform notation of tableaux rules for multiple-valued logics. *Pages 238–245 of: Proc. International Symposium on Multiple-Valued Logic, Victoria.* IEEE Press.

Hähnle, R. (1992a). A new translation from deduction into integer programming. *In: Proc. Conf. on Artificial Intelligence and Symbolic Mathematical Computations, Karlsruhe.* Springer LNCS.

Hähnle, R. (1992b). Tableaux-Based Theorem Proving in Multiple-Valued Logics. Ph.D. thesis, University of Karlsruhe, Department of Computer Science.

Hähnle, R. (1993a). Many-valued logic and mixed integer programming. Forthcoming.

Hähnle, R. (1993b). Short normal forms for arbitrary finitely-valued logics. *In: Proceedings ISMIS'93, Trondheim, Norway.* Springer LNCS.

Hähnle, R. (1993c). Uses of many-valued deduction in hardware verification. *In: Proc. ITG/GI Workshop Formale Methoden zum Entwurf korrekter Systeme, Bad Herrenalb.*

Hähnle, R., and Kernig, W. (1993). Verification of switch level designs with many-valued logic. *In: Proc. LPAR'93, St. Petersburg (to appear).* Springer, LNAI 689.

Hähnle, R., and Schmitt, P. H. (1993). The liberalized δ-rule in free variable semantic tableaux. *Journal of Automated Reasoning, to appear.*

Hähnle, R., Beckert, B., Gerberding, S., and Kernig, W. (1992). *The Many-Valued Tableau-Based Theorem Prover $_3T^AP$.* IWBS Report 227. Wissenschaftliches Zentrum Heidelberg, IWBS, IBM Deutschland.

Hähnle, R., Lu, J. J., Murray, N. V., and Rosenthal, E. (1993). *Many-Valued Resolution on Signed Clauses (tentative).* In preparation.

Hayes, J. P. (1982). A unified switching theory with applications to VLSI design. *Proceedings of the IEEE,* **70**(10), 1140–1151.

Hayes, J. P. (1986). Pseudo-Boolean logic circuits. *IEEE Transactions on Computers,* **C-35**(7), 602–612.

Hintikka, K. J. J. (1955). Form and content in quantification theory. *Acta Philosohica Fennica,* **8**, 7–55.

Hooker, J. N. (1988). A quantitative approach to logical inference. *Decision Support Systems,* **4**, 45–69.

Hooker, J. N. (1991). Logical inference and polyhedral projection. *In: Proc. Computer Science Logic Workshop 1991, Berne.* Springer, LNCS.

Hooker, J. N., and Fedjki, C. (1990). Branch-and-cut solution of inference problems in propositional logic. *Annals of Mathematics and Artificial Intelligence,* **1**, 123–139.

Hughes, G. E., and Cresswell, M. J. (1984). *A Companion to Modal Logic.* Methuen Press.

ISMVL-21. (1991). *21st International Symposium on Multiple-Valued Logic, Victoria, Canada.* IEEE Press.

ISMVL-22. (1992). *22nd International Symposium on Multiple-Valued Logic, Japan.* IEEE Press.

Janssen, G. (1989). Hardware verification using temporal logic: A practical view. *Pages 291–300 of: Proceedings Workshop Applied Formal Methods for correct VLSI Design, Leuwen.*

Jeroslow, R. G. (1988). *Logic-Based Decision Support. Mixed Integer Model Formulation.* Elsevier, Amsterdam.

Jeroslow, R. G., and Wang, J. (1990). Solving propositional satisfiability problems. *Annals of Mathematics and Artificial Intelligence,* **1**, 167–187.

Kapetanović, M., and Krapež, A. (1989). A proof procedure for the first-order logic. *Publications de l'Institut Mathematiqu, nouvelle série,*

59(45), 3–5.

Karnaugh, M. (1953). The map method for synthesis of combinational logic circuits. *AIEE Transactions, Part I Communication and Electronics,* **72**(November), 593–599.

Karp, R. M. (1972). Reducibility among combinatorial problems. *Pages 85–103 of:* Miller, R., and Thatcher, J. (eds.), *Complexity of Computer Computations.* Plenum Press.

Kenevan, J. R., and Neapolitan, R. E. (1992). A model theoretic approach to propositional fuzzy logic using Beth tableaux. *Pages 141–158 of:* Zadeh, L. A., and Kacprzyk, J. (eds.), *Fuzzy Logic for the Management of Uncertainty.* John Wiley & Sons.

Kernig, W. (1992). Modellierung und Verifikation von Switch-Level Spezifikationen mit Hilfe von mehrwertiger Logik. Master's thesis, University of Karlsruhe, Department of Computer Science.

Kleene, S. (1938). On a notation for ordinal numbers. *Journal of Symbolic Logic,* **3**, 150–155.

Kunen, K. (1987). Negation in logic programming. *Journal of Logic Programming,* **4**, 289–308.

Langholm, T. (1989). Algorithms for Partial Logic. Unpublished manuscript.

Lee, C. Y. (1959). Representation of switching circuits by binary-decision programs. *Bell System Technical Journal,* **38**(July), 985–999.

Lee, R. C. T. (1972). Fuzzy logic and the resolution principle. *Journal of the ACM,* **19**(1), 109–119.

Lee, R. C. T., and Chang, C.-L. (1971). Some properties of fuzzy logic. *Information and Control,* **19**(5), 417–431.

Lee, S.-J., and Plaisted, D. A. (1992). Eliminating duplication with the hyper-linking strategy. *Journal of Automated Reasoning,* **9**(1), 25–42.

Letz, R. (1993). First-Order Calculi and Proof Procedures for Automated Deduction. Ph.D. thesis, TU Darmstadt.

Letz, R., Schumann, J., Bayerl, S., and Bibel, W. (1991). *SETHEO: A High-Perfomance Theorem Prover.* Technical report. Forschungsgruppe Künstliche Intelligenz, TU München.

Lloyd, J. W. (1987). *Foundations of Logic Programming.* Second edn. Springer, Berlin.

Łoś, J., and Suszko, R. (1958). Remarks on sentential logics. *Indigationes Mathematicae,* **20**, 177–183.

Loveland, D. (1969). A simplified format for the model elimination procedure. *Journal of the ACM,* **16**(3), 233–248.

Lu, J. J., Henschen, L. J., Subrahmanian, V., and da Costa, N. C. A. (1991). Reasoning in paraconsistent logics. *Pages 181–210 of:* Boyer, R. (ed.), *Automated Reasoning: Essays in Honor of Woody Bledsoe.* Kluwer.

Lu, J. J., Murray, N. V., and Rosenthal, E. (1993). *Signed Formulas and*

Annotated Logics. Technical report. Bucknell University, Lewisburg, PA.

Łukasiewicz, J. (1920). O logice tròjwartościowej. *Ruch Filozoficzny*, **5**, 169–171.

MacNish, C., and Fallside, F. (1990). *Asserted 3-valued Logic for Default Reasoning.* CUED/F-INFENG/TR.40. Department of Engineering, University of Cambridge.

Manna, Z., and Waldinger, R. (1986). Special relations in automated deduction. *Journal of the ACM*, **33**(1), 1–59.

Manthey, R., and Bry, F. (1988). SATCHMO: A theorem prover implemented in Prolog. *Pages 415–434 of: Proceedings 9th Conference on Automated Deduction.* Springer LNCS, New York.

McNaughton, R. (1951). A theorem about infinite-valued sentential logic. *Journal of Symbolic Logic*, **16**(1), 1–13.

McRobbie, M. A., Meyer, R. K., and Thistlewaite, P. B. (1988). Towards efficient 'knowledge-based' automated theorem proving for non-standard logics. *Pages 197–217 of: Proceedings 9th CADE.* LNCS. Springer, New York.

Mellouli, T. (1990). A Tree Representation of the Modified Problem Reduction Format and its Extension to Three-Valued Logic. Ext. Version of talk given at German-Japanese Workshop on Logic and Natural Language, Kyoto.

Mints, G. (1990). Gentzen-type systems and resolution rules, part 1: Propositional logic. *Pages 198–231 of: Proc. COLOG-88, Tallin.* LNCS, vol. 417. Springer.

Mondadori, M. (1988). *Classical Analytical Deduction.* Annali dell' Università di Ferrara, Nuova Serie, sezione III, Filosofia, discussion paper, n. 1. Università degli Studi di Ferrara.

Mondadori, M. (1989). *Classical Analytical Deduction, Part II.* Annali dell' Università di Ferrara, Nuova Serie, sezione III, Filosofia, discussion paper, n. 5. Università degli Studi di Ferrara.

Morgan, C. G. (1976). A resolution principle for a class of many-valued logics. *Logique et Analyse*, **19**(74–75–76), 311–339.

Morgan, C. G. (1985). Autologic. *Logique et Analyse*, **28**(110–111), 257–282.

Mostowski, A. (1957). On a generalization of quantifiers. *Fundamenta Mathematicæ*, **XLIV**, 12–36.

Mundici, D. (1986). Interpretation of AF C*-algebras in Łukasiewicz sentential calculus. *Journal of Functional Analysis*, **65**, 15–63.

Mundici, D. (1987a). Satisfiability in many-valued sentential logic is NP-complete. *Theoretical Computer Science*, **52**, 145–153.

Mundici, D. (1987b). The Turing complexity of AF C*-algebras with lattice-ordered Ko. *Pages 256–264 of:* Börger, E. (ed.), *Computation Theory and Logic.* LNCS, vol. 270. Springer, Heidelberg.

Mundici, D. (1989). The logic of Ulam's game with lies. *In: Proceedings International Conference Knowledge, Belief and Strategic Interaction, Castiglioncello.* Cambridge Studies Series in Probability, Induction and Decision Theory.

Mundici, D. (1990). The complexity of adaptive error-correcting codes. *Pages 300–307 of: Proceedings Workshop Computer Science Logic 90, Heidelberg.* Springer, LNCS 533.

Mundici, D. (1991). Normal forms in infinite-valued logic: The case of one variable. *In: Proceedings Workshop Computer Science Logic 91, Berne.* Springer, LNCS.

Mundici, D. (1993). A constructive proof of McNaughton's Theorem in infinite-valued logic. Forthcoming.

Murray, N. V. (1982). Completely non-clausal theorem proving. *Artificial Intelligence*, **18**, 67–85.

Murray, N. V., and Rosenthal, E. (1986). *Path Dissolution for Propositional Logic.* Technical Report TR-86-6. Department of Computer Science, SUNY at Albany.

Murray, N. V., and Rosenthal, E. (1987). Path dissolution: a strongly complete rule of inference. *Pages 161–166 of: Proc. of the 6th National Conference on Artificial Intelligence, Seattle.*

Murray, N. V., and Rosenthal, E. (1990a). *DISSOLUTION: Making paths vanish.* Technical Report TR-90-?? Department of Computer Science, SUNY at Albany.

Murray, N. V., and Rosenthal, E. (1990b). *On the Relative Merits of Path Dissolution and the Method of Analytical Tableaux.* Technical Report TR-90-5. Department of Computer Science, SUNY at Albany.

Murray, N. V., and Rosenthal, E. (1991a). Improving tableau deductions in multiple-valued logics. *Pages 230–237 of: Proceedings 21st International Symposium on Multiple-Valued Logic, Victoria.* IEEE Computer Society Press, Los Alamitos.

Murray, N. V., and Rosenthal, E. (1991b). Resolution and path-dissolution in multiple-valued logics. *In: Proceedings International Symposium on Methodologies for Intelligent Systems, Charlotte.*

Murray, N. V., and Rosenthal, E. (1991c). *Signed Formulas: A Classical Approach to Multiple-Valued Logics.* Technical Report TR 91-12. SUNY at Albany, NY.

Murray, N. V., and Rosenthal, E. (1993a). Making paths vanish. To appear in *Journal of the ACM.*

Murray, N. V., and Rosenthal, E. (1993b). Signed formulas: A liftable meta logic for multiple-valued logics. *In: Proceedings ISMIS'93, Trondheim, Norway.* Springer LNCS.

Nemhauser, G. L., and Wolsey, L. A. (1989). Integer programing. *Chap. VI, pages 447–527 of:* Nemhauser, G. L., Kan, A. H. G. R., and Todd, M. J. (eds.), *Handbooks in Operations Research and Manage-*

ment Science, Vol. II: Optimization. Amsterdam: North-Holland.

Nipkow, T. (1988). Unification in primal algebras. *In: Proc. 13th Colloquium on Trees in Algebra and Programming.* Springer, LNCS 299.

O'Hearn, P., and Stachniak, Z. (1992). A resolution framework for finitely-valued first-order logics. *Journal of Symbolic Computing,* **13**, 235–254.

Ohlbach, H. J. (1989). *Context Logic.* Technical Report SEKI Report SR-89-08. University of Kaiserslautern.

Ophelders, W. M. J. (1992). Automated Theorem Proving Based upon a Tableau-Method with Unification under Restrictions. Ph.D. thesis, University of Tilburg.

Oppacher, F., and Suen, E. (1988). HARP: A tableau-based theorem prover. *Journal of Automated Reasoning,* **4**, 69–100.

Orłowska, E. (1967). Mechanical proof procedure for the n-valued propositional calculus. *Bull. de L'Acad. Pol. des Sci., Série des sci. math., astr. et phys.,* **XV**(8), 537–541.

Orłowska, E. (1969a). Automatic theorem proving in a certain class of formulae of predicate calculus. *Bull. de L'Acad. Pol. des Sci., Série des sci. math., astr. et phys.,* **XVII**(3), 117–119.

Orłowska, E. (1969b). Mechanical theorem proving in a certain class of formulae of the predicate calculus. *Studia Logica,* **XXV**, 17–27.

Orłowska, E. (1978). The resolution principle for ω^+-valued logic. *Fundamenta Informaticae,* **II**(1), 1–15.

Orłowska, E. (1980). Resolution systems and their applications II. *Fundamenta Informaticae,* **III**(3), 333–362.

Orłowska, E. (1985). Mechanical proof methods for Post Logics. *Logique et Analyse,* **28**(110), 173–192.

Patel-Schneider, P. F. (1989). A four-valued semantics for terminological logics. *Artificial Intelligence,* **38**, 319–351.

Patel-Schneider, P. F. (1990). A decidable first-order logic for knowledge representation. *Journal of Automated Reasoning,* **6**, 361–388.

Plaisted, D. A. (1988). Non-Horn clause logic programming without contrapositives. *Journal of Automated Reasoning,* **4**, 287–325.

Plaisted, D. A., and Lee, S.-J. (1990). Inference by clause matching. *Chap. 8, pages 200–235 of:* Ras, Z., and Zemankova, M. (eds.), *Intelligent Systems—State of the art and future directions.* Ellis Horwood.

Posegga, J. (1992a). Deduction based on Shannon graphs. *In: Proceedings GWAI-92, Bonn.* Springer, LNCS.

Posegga, J. (1992b). First-order shannon graphs. *In: Intern. Conf. on Fifth Generation Computer Systems/Workshop on Automated Deduction.* TM-1184. ICOT, Tokyo, Japan.

Posegga, J. (1993). First-order Deduction with Shannon Graphs. Ph.D. thesis, University of Karlsruhe.

Post, E. L. (1921). Introduction to a general theory of elementary propositions. *Pages 264–283 of:* van Heijenoort, J. (ed.), *From Frege to*

Gödel. A Source Book in Mathematical Logic, 1879–1931. Harvard University Press, Cambridge MA.

Prawitz, D. (1970). A proof procedure with matrix reduction. *Pages 207–213 of: LNM 125.* Springer.

Quine, W. V. (1952). The problem of simplifying truth functions. *American Mathematical Monthly*, **59**(oct), 521–531.

Rasiowa, H. (1973). On generalized Post algebras of order ω^+ and ω^+-valued predicate calculi. *Bull. Acad. Polon. Sci., Sèrie Sci. Math. Astr. Phys.*, **XXI**, 209–219.

Rasiowa, H. (1974). *An Algebraic Approach to Non-Classical Logics.* Studies in Logic and the Foundations of Mathematics, vol. 78. North-Holland, Amsterdam.

Reeves, S. V. (1987). Semantic Tableaux as a Framework for Automated Theorem-Proving. Department of Computer Science and Statistics, Queen Mary College, Univ. of London.

Rescher, N. (1969). *Many-Valued Logic.* McGraw-Hill, New York.

Ries, K., and Hähnle, R. (1993). Prädikatenlogisches Beweisen mit gemischt ganzzahliger Optimierung. Ein tableaubasierter Ansatz. *In: Working Notes of Workshop Künstliche Intelligenz und Operations Research, Berlin.*

Rosser, J. B., and Turquette, A. R. (1952). *Many-Valued Logics.* Amsterdam: North-Holland.

Rousseau, G. (1967). Sequents in many valued logic I. *Fundamenta Mathematicæ*, **LX**, 23–33.

Salkin, H., and Mathur, K. (1989). *Foundations of Integer Programming.* North-Holland.

Schmitt, P. H. (1986). Computational aspects of three-valued logic. *Pages 190–198 of:* Siekmann, J. H. (ed.), *Proc. 8th International Conference on Automated Deduction.* Springer, LNCS.

Schmitt, P. H. (1990). *Logik (Nichtklassische Logiken).* Vorlesungsskript Universität Karlsruhe.

Schöpke, G. (1990). Wissenserwerb für Expertensysteme durch computerunterstützte logisch-mathematische Modellierung. *In: Proc. 5. Symposium Grundlegung und Anwendung der Informatik, Chemnitz.*

Schöpke, G. (1991). *Möglichkeiten des Einsatzes eines dreiwertigen Theorembeweisers.* IWBS Report 188. Wissenschaftliches Zentrum, IWBS, IBM Deutschland.

Schröter, K. (1955). Methoden zur Axiomatisierung beliebiger Aussagen- und Prädikatenkalküle. *Zeitschrift für math. Logik und Grundlagen der Mathematik*, **1**, 241–251.

Schuhmann, J., Trapp, N., and van der Koelen, M. (1991). *SETHEO—Users Manual.* Forschungsgruppe Künstliche Intelligenz, TU München.

Schwind, C. B. (1990). A tableaux-based theorem prover for a decidable subset of default logic. *Pages 528–542 of: Proc. 10th International Conference on Automated Deduction, Kaiserslautern.* Springer, LNCS 449.

Schwind, C. B., and Risch, V. (1991). A tableaux-based characterisation for default logic. *In: European Conference on Symbolic and Quantitative Approaches for Uncertainty.*

Scott, D. (1976). Does many-valued logic have any use? *Pages 64–74 of:* Körner, S. (ed.), *Philosophy of Logic.* Blackwell, Oxford.

Seuren, P. A. M. (1985). *Discourse Semantics.* Blackwell, Oxford.

Shannon, C. E. (1938). A symbolic analysis of relay and switching circuits. *AIEE Transactions,* **67**, 713–723.

Sheperdson, J. C. (1989). A sound and complete semantics for a version of negation as failure. *Theoretical Computer Science,* **65**, 343–371.

Smullyan, R. (1968). *First-Order Logic.* Springer, New York.

Stachniak, Z. (1988). The resolution rule: An algebraic perspective. *Pages 227–242 of: Proc. of Algebraic Logic and Universal Algebra in Computer Science Conf.* Springer LNCS 425, Heidelberg.

Stachniak, Z. (1990a). Note on effective constructability of resolution proof systems. *Pages 487–498 of: Proc. of Logic in AI European Workshop, Amsterdam.* Lecture Notes in AI, vol. 478. Springer.

Stachniak, Z. (1990b). Note on resolution approximation of many-valued logics. *Pages 204–209 of: 20th International Symposium on Multiple-Valued Logic, Charlotte.* IEEE Press.

Stachniak, Z. (1991a). Extending resolution to resolution logics. *Journal for Experimental and Theoretical Artificial Intelligence,* **3**, 17–32.

Stachniak, Z. (1991b). Minimization of resolution proof systems. *Fundamenta Informaticae,* **XIV**, 129–146.

Stachniak, Z. (1991c). Note on resolution circuits. *In: Proc. Intern. Symp. on Methodologies for Intelligent Systems, Charlotte.* Springer, LNAI.

Stachniak, Z. (1992). Resolution approximation of first-order logics. *Information and Computation,* **96**(2), 225–244.

Stachniak, Z. (1993). Algebraic semantics for cumulative inference operations. *In: Proc. AAAI'93, Raleigh/NC.*

Stachniak, Z., and O'Hearn, P. (1990). Resolution in the domain of strongly finite logics. *Fundamenta Informaticae,* **XIII**, 333–351.

Stickel, M. E. (1988). A prolog technology theorem prover. *Pages 752–753 of:* Lusk, E., and Overbeek, R. (eds.), *Proc. 9th International Conference on Automated Deduction.* Springer LNCS, New York.

Stickel, M. E. (1992). A Prolog technology theorem prover: A new exposition and implementation in Prolog. *Theoretical Computer Science,* **104**(1), 109–129.

Suchoń, W. (1974). La méthode de Smullyan de construire le calcul n-valent de Łukasiewicz avec implication et négation. *Reports on Mathematical Logic, Universities of Cracow and Katowice,* **2**, 37–42.

Surma, S. J. (1984). An algorithm for axiomatizing every finite logic. *Pages 143–149 of:* Rine, D. C. (ed.), *Computer Science and Multiple-Valued Logics.* North-Holland, Amsterdam.

Tseitin, G. (1970). On the complexity of proofs in propositional logics. *Seminars in Mathematics,* **8.** Reprinted in: Siekmann, J. and Wrightson G. (eds.) *Automation of Reasoning 2: Classical Papers on Computational Logic,* pp. 466–483, Springer, 1983.

Urquhart, A. (1986). Many-valued logic. *Chap. 2, pages 71–116 of:* Gabbay, D., and Guenthner, F. (eds.), *Handbook of Philosophical Logic, Vol. III: Alternatives in Classical Logic.* Reidel, Dordrecht.

Wallen, L. A. (1990). *Automated Proof Search in Non-Classical Logics.* MIT Press.

Weigert, T. J., Tsai, J.-P., and Lu, X. (1993). Fuzzy operator logic and fuzzy resolution. *Journal of Automated Reasoning,* **10,** 59–78.

Wójcicki, R. (1988). *Theory of Logical Calculi.* Reidel, Dordrecht.

Wolf, R. G. (1977). A critical survey of many-valued logics 1966–1974. *Pages 468–474 of:* Dunn, J. M., and Epstein, G. (eds.), *Modern Uses of Multiple-Valued Logic.* Reidel, Dordrecht.

Wolper, P. (1981). Temporal logic can be more expressive. *Pages 340–348 of: Proceedings 22nd Annual Symposium on Foundations of Computer Science.*

Wos, L. (1988). *Automated Reasoning: 33 Basic Research Problems.* Prentice-Hall, Englewood Cliffs.

Wos, L., Overbeek, R., Lusk, E., and Boyle, J. (1984). *Automated Reasoning—Introduction and Applications.* Prentice-Hall, Englewood Cliffs.

Yurchak, J. M., and Butler, J. T. (1990). *HAMLET User Reference Manual.* 4th edn. Naval Postgraduate School.

Zach, R. (1992). *A Many-Valued Logic for Default Reasoning.* Technical Report TR-E185.2-Z-2-1992. Technical University of Wien, Institut für Computersprachen.

Zadeh, L. A. (1965). Fuzzy sets. *Information and Control,* **8,** 338–353.

Zadeh, L. A. (1988). Fuzzy logic. *Computer,* **21**(4).

INDEX

Underlined page numbers indicate the place, where a concept is defined. Page numbers in italics indicate occurrences in footnotes.

$\mathcal{L}_3$, <u>125</u>
$\mathcal{L}_M^n$, <u>12</u>, 24, 49, 122, 124
$\mathcal{L}_{M+}^n$, <u>12</u>
$\mathcal{L}_{M+}^3$, 12, 33, 40, 105, 124
$\mathcal{L}_{EP}^n$, <u>135</u>
$\mathcal{L}_{EP}^{(n,m)}$, <u>135</u>
$\mathcal{L}_P^n$, <u>14</u>, 69
$\mathcal{L}_{Sm}^n$, <u>56</u>, 59
$\mathcal{L}_\omega$, <u>14</u>, 99, 100, <u>147</u>
$\mathcal{L}_{UNF}^n$, 145
$\mathcal{L}_{SKL}^n$, <u>13</u>, 59
$_3\mathcal{T}^4\!P$, 108, 118

abstract algebra, <u>4</u>, 5, 119, 142
    free, <u>5</u>, *5*
    free generators of, <u>5</u>
    free term algebra, <u>7</u>
    generators, <u>5</u>
    similar, <u>4</u>
    subalgebra, <u>4</u>
    unification in, 142
    universe of, <u>4</u>
ACP, *see* analytic consistency property
algebra of signs, <u>33</u>
analytic consistency property, <u>44</u>, 47
analytic tableau, *see* tableau
assignment, <u>9</u>
    variant of, <u>9</u>
atomic formula
    first-order, <u>8</u>
AUTOLOGIC, 151, 153

BDD, *see* decision diagram
bilattice, 113
binary decision diagram, *see* decision diagram
bMIP-representable, <u>90</u>, 91

circle, 60, <u>60</u>

clause, <u>11</u>
    definite Horn, *149*
    Horn, *148*
CNF, *see* normal form, conjunctive
compilation of logic problems, 102
complement closed, <u>87</u>
completeness, 19, 27, 42–47, 63, 66, 68, 73, 86, 88, 94, 124, 126, 138, 142
    strong, 21
complexity
    of proofs, 36, 48, 83, 85, 95, 147
    of satisfiability problem, 100, 148
conclusion, <u>34</u>
connection method, 144
connective, <u>5</u>
    0-ary, 47
    as filter, 78
    conjugate, <u>66</u>
    many-valued generalization, <u>7</u>, *13*, 56
    primary, <u>55</u>, 55–59, 62
        many-valued, 56
    regular, <u>60</u>, 61, 80
        direction of, <u>60</u>
        number of, 70
        orientation of, <u>60</u>
        starting point of, <u>60</u>
        threshold of, <u>61</u>, 65, <u>66</u>
    strong, <u>7</u>, 12, 13
    weak, <u>7</u>, 12, 13
consequence, <u>10</u>, 21
consequence relation, <u>128</u>
    strongly finite, <u>128</u>
consistent, <u>44</u>
constraint rule, <u>92</u>
    for modal logic, 116
constraint tableau proof, <u>94</u>
contraction rule, <u>76</u>
contradiction set, <u>19</u>, 24, <u>36</u>, 47
control variable, <u>99</u>
corner set, <u>60</u>
criteria for theorem provers, 2–3, 107–108, 153
cut, 85

decision diagram, 102–104, 119, 139, 153
deduction theorem, 21
depth of formula, <u>6</u>, 8
dissolution, 104, 105, 144, 153

many-valued, 146, 147
distribution quantifier, 27, 71, 143
tableau rule for, 29
downset, 79, 138

error-correcting codes, 112
extension, 34
intersection of, 85

fairness, 20, 86
falsified, 134
finite character, 44
first-order
assignment, 9
atomic formula, 8
consequence, 10
formula, 8
language, 8
logic, 9, 102
satisfiable, 10
structure, 9
valuation, 10
formula
abbreviation, 10, 122
atomic, 5, 19
closed, 8
complement, 87
complementary, 18, 19, 24, 36,
63, 137
distribution of, 26
first-order, 8
logical equivalent, 122, 123
propositional, 5, 6
refutational equivalent, 122,
123
schema, 10
self-contradictory, 24, 36
signed, 17, 41, 105
syntactical equivalent, 11
syntactical variant, 11
used, 20
function minimization, 50–54, 87,
122
functionally complete, 14, 69, 119,
126, 143, 153
fuzzy logic, 80, 121, 133, *133*

Gentzen system, 143

hardware verification, 112, 117–119
Herbrand
structure, 73, 76

theorem, 124
Hilbert system, 111, 112, 117, *132*,
151, *151*
Hintikka
lemma, 43
set, 42, 44, 47, 88
saturated, 43
homomorphism, 5, 34, 35, 41, 46

implicant, 51
prime, 51, 52
essential, 51
independence of axioms, 111, 112
integer programming, 90, 93, 95,
102
interval arithmetic, 112, 117
IP, *see* integer programming

Karnaugh map, 50, 51, 86
KE system, 83, 84
analytic, 84
many-valued, 87, 88
unrestricted, 84
knowledge representation, 111

labelled deductive system, 22, 151,
153
lemma generation, 36, 83–87
Lindenbaum construction, 44
linear programming, 90, 102, 148
link, 145
partial, 146
literal, 11
operator, 50
signed, 146
logic
classical, 16, 58
default, 111
functionally complete, 12, 13,
121
modal, 111, 116
non-monotonic, 111
nonclassical, 102
pseudo-regular, 78
regular, 63, 59–77, 137
relevant, 111, 113, 114
logic programming, 110, 113, 148
LP, *see* linear programming
Łukasiewicz logic, 13, 49, 80, 92, 94,
98, 115, 148
infinitely valued, 99, 112

many-valued logic, 1

applications of, 110–119
bibliography, 120
complexity of, 31
in hardware verification, 118
lemma generation in, 85
non-linear, 80, 113
use of, 110
matrix, *see* propositional matrix
McNaughton
function, 147
theorem, 100, 147
metric, 60, *60*
minterm, 50
MIP, *see* mixed integer programming
mixed integer programming, 89, 90, 91, *94*, 97, 99, 100, 105
model, 7
model existence theorem, 44
model set, *see* Hintikka set
modified problem reduction format, 148, 149
multi-valued logic, *see* many-valued logic
multiple-valued logic, *see* many-valued logic

natural language processing, 111
normal form, 121, 122, 129
conjunctive, 11
D-CNF, 137
D-DNF, 137
J-CNF, 122
negation, 11, 144, 145
prenex, 11
CNF, 11
existential, 11
matrix, 11
universal, 11
S-CNF, 126
unary, 104, 145

ON-set, 50, 52
operator, *see* propositional operator

paraconsistent logic, 138
parameter, 8
path dissolution, *see* dissolution
poset, 79
Post logic, 14, 69
extended, 135
generalized, 136

prefix, *see* sign
premise, 34
principle of bivalence, 84, 84
restricted, 84
principle of multivalence, 88
product term, 51
program verification, 110
Prolog, 97, 148, 149
propositional
connective, 5
formula, 6
language, 6
logic, 6, 22
matrix, 6, 113
valuation, 6
variable, 5

quantification, 9
many-valued, 27, 71–77, 103, 143

resolution, 121–139
approximation, 131
hyper-, 123
proof, 124
rule, 122
Morgan, 123
non-clausal, 129
Orłowska, 137
Schmitt, 126
Stachniak, 129
rule, *see* tableau or resolution

satisfiable, 7, 10, 122
tableau, 41
selection rule, 85, 86
semantic complement, 85
semantic tableau, *see* tableau
sentence, *see* formula, closed
sets-as-signs, 2, 33, 40, 71, 80, 85, 104, 109, 116, 117, 119, 154
Shannon graph, *see* decision diagram
sign, 22–23, 33
algebra of ~s, 33
base set of ~s, 33
classical, 17
complement, 87
complementary, 18
complete set of ~s, 40
corresponding to Kripke model, 116

in regular logic, 62, 63, 76, 80,
        91, 138
many-valued, 23
similarity type, 4
size, *see* complexity
Skolem symbol, 8, 17
SOP, *see* sum-of-products
soundness, 19, 40–42, 63, 66, 68, 73,
        86, 94, 124, 126, 138, 141
    strong, 21
splitting rule, 76
subset closed, 44
substitution, 8
sum-of-products, 50, 51, 104
    minimal ∼ representation, 51
switch level model, 118

tableau, 17, 15–23, 139, 143, 147
    branch, 18
        closed, 18, 19, 94
        complete, 19
        in focus, 82
        open, 18, 94
    closed, 19, 94, 138
    complete, 19
    construction, 17, 94
    free variable, 20, 103
    ground, 20
    number of necessary trees, 25,
        29
    proof, 18
        redundancy in, 30, 32
        with constraints, 93
    rule, 34, 53
        asymmetric, 82, 84
        branching factor, 24
        conclusion, 34
        constrained, 92
        extension, 34
        finding of, 53
        for quantified formulas, 29,
            73, 77
        for regular logic, 63–66, 68
        incomplete, 27
        many-valued, 24
        number of extensions, 24, 30,
            34, 48, 49
        number of formulas, 30, 35
        premise, 34
        proviso, 92
        uniqueness of, 35, 50, 53, 81
        with lemma generation, 85

satisfiable, 41
systematic, 20
tautology
    first-order, 10
    propositional, 7
term, 7
truth maintenance system, 111
truth value, 6
    conjugate, 55
    designated, 6, 6, 57
    order, 78, 79, 113

UNF, *see* normal form, unary
unification, 142
unified notation, *see* uniform notation
uniform notation, 16–18, 24, 55–80,
        137
unsatisfiable, 134
upset, 79, 138

valuation, 9, 10
    first-order, 10
    propositional, 6
variable
    bound, 8
    free, 8, 103
verifier, 129, 131
    number of, 131